# GLOBAL SUSTAINABILITY

First Edition

Edited by Mark White and Phoebe Crisman
*University of Virginia*

Bassim Hamadeh, CEO and Publisher
Jamie Giganti, Managing Editor
Jess Busch, Graphic Design Supervisor
Melissa Barcomb, Acquisitions Editor
Sarah Wheeler, Senior Project Editor
Stephanie Sandler, Licensing Associate

Printed in the United States of America

ISBN: 978-1-62131-977-1 (pbk)

www.cognella.com    800-200-3908

# CONTENTS

## SECTION III: VISION AND COMMUNITY ENGAGEMENT

## SECTION IV: ETHICS, LAW AND POLICY

## SECTION V: WATER, ENERGY, FOOD AND HEALTH

# SECTION VI: COMMUNITIES AND THE BUILT ENVIRONMENT

# SECTION VII: COMMERCE AND PRODUCTION

# SECTION VIII: PERSONAL BEHAVIOR CHOICES

# FOREWORD

By Mark White and Phoebe Crisman

The readings included in **Global Sustainability** provide an introduction to the concept of *sustainability*, broadly defined as a "society of permanence," a world in which humanity ensures its well-being across the generations by improving the stability of ecological and sociocultural systems. Achieving a sustainable society will require decision makers to address issues of scale, just distribution, and economic efficiency.

Human societies have developed numerous place-based solutions addressing their common needs of food, clothing, shelter, belonging, and self-actualization. Unfortunately, as we have adapted local environments to meet our needs, wants, and desires, we have often reduced the ability of natural ecosystems to meet these demands in the future. Moreover, in an increasingly globalized world, choices and actions taken in one region can have unintended consequences for populations and environments far distant from their source.

Section I, **Introduction and the State of the Planet**, introduces the central challenge of sustainability—increasing demand and finite resources.[1] In his essay "The Problem of Production," economist E. F. Schumacher argues that our failure to differentiate between income and capital is at the heart of the matter. "From Muir to Meadows" by Simon Dresner provides a brief historical overview of key efforts to recognize and address the issue of limited resources. Paul Conkin's contribution, "Population, Consumption, and the Environment," highlights the role of population growth, observing that the pressures we're facing derive from two sources, "the multiplying poor and the indulgent rich," after which Bill McKibben, in "A New World," eloquently describes the many ways in

---

1   Ecological footprint proponent Mathis Wackernagel's tongue-in-cheek definition of *sustainability* is "one planet."

which global climate change is altering human activities on the planet.

Section II, **Systems Thinking and the Limits to Growth**, begins with Donella Meadows's brief introduction to systems thinking, "The System Lens." She shares the lessons that everything is connected to everything else, and that deeper system structures—connections, feedback loops, and delays—must be better understood if we are to advance toward achieving a sustainable future. Nørgaard, Peet and Ragnardóttir's article, "The History of The Limits to Growth" recounts the impact and subsequent debate surrounding the seminal 1972 *Limits to Growth* study, which applies these lessons to the challenge of sustainability with somewhat sobering—yet realistic—results.

Section III, **Vision and Community Engagement**, comprises two readings: Ernest Boyer's classic essay "The Scholarship of Engagement" and Mark Woods's "From Service to Solidarity." While Boyer reviews and applauds the role higher education has played in building a more just and civil society, he encourages an even closer connection between academic and civic cultures as a means of improving overall human well-being. Acknowledging and echoing Boyer's sentiments, Woods exhorts teachers to involve students more deeply in social, economic, and moral problems, with the goal of producing "transformed nonconformists."

Section IV, **Ethics, Law, and Society**, provides a short introduction to the ethical and legal issues associated with achieving a sustainable future. Tom Russ's reading, "Is There an Ethical Obligation to Act Sustainably?" describes several theories of ethics, with an especial focus on the environment and sustainability. In "Environmental Laws and Sustainability," John Dernbach and Joel Mintz note that achieving a sustainable society will most certainly require new laws and institutions, and then suggest a

list of ten lessons about what a "law for sustainability" might involve.

Section V, **Water, Energy, Food, and Health**, contains five readings addressing these important challenges. In "Water: Adapting to a New Normal," Sandra Postel reviews the history and current state of our water challenges, ultimately calling for better policy and better pricing of Earth's aquatic resources. Fereidoon Sioshansi's chapter, "Why Do We Use So Much Energy, and What For?" describes the myriad ways in which we use energy and questions whether we might be able to achieve equally satisfying—or potentially superior—standards of living with less energy consumption than we're currently using. Anna Lappé's "The Climate Crisis at the End of Your Fork" explores the link between food production and climate change. According to Lappé, deforestation, fertilizer production, livestock emissions, and energy associated with transporting foodstuffs from farm to market are responsible for up to one-third of total global warming effects. In "The Pleasures of Eating," farmer-philosopher Wendell Berry argues that responsible eating is not only beneficial and prudent but also gastronomically fulfilling. Finally, Cindy Parker and Brian Schwartz describe a number of looming health challenges, including heat stress and worsening air quality, infectious diseases, and threatened water and food supplies in their article, "Human Health and Well-Being in an Era of Energy Scarcity and Climate Change."

Section VI, **Communities and the Built Environment**, addresses a key aspect of humanity's planetary footprint—where and how we build. Tim Beatley and Kristy Manning open this section with an investigation of what makes a community "sustainable." Among the characteristics noted by the authors are an appreciation of limits, social equity, integration with the natural environment, and full-cost accounting. Alex Krieger, in "The Costs—or Have There

Been Benefits, Too?—of Sprawl," continues this examination with respect to urban sprawl, ultimately concluding that sprawl seems to result from a combination of private benefits and public costs. Peter Buchanan urges the immediate adoption of "green" building techniques to reduce energy demand (both embodied and operating) and provides several examples of good design. Finally, noting that "buildings are the single most important contributor to the greenhouse gas emissions that cause climate change," the Pew Charitable Trust's primer "Building Solutions to Climate Change" reviews the challenge and current efforts to mitigate the impact of adverse climate conditions.

Section VII, **Commerce and Production**, opens with Hawken, Lovins and Lovins's indictment of industrial capitalism in their article "The Next Industrial Revolution." The authors emphasize the importance of *natural capital,* defined as the benefits humans receive from nature, and argue strongly for a restorative commerce that takes limits seriously. To some extent, these issues are addressed in "The Business Case for the Green Economy," prepared by the United Nations Environment Programme, which outlines the business case for sustainability and highlights numerous benefits (e.g., increased sales, reduced capital expenditures, improved profit margins, reduced cost of capital) accruing to proactive businesses.

Section VIII, **Personal Behavior Choices**, concludes the book with James Farrell's "Making Environmental History," targeted at students living on college campuses. Ultimately, we all need to "walk the talk" by explicitly examining our own environmental and social impacts with the goal of minimizing our contribution to humanity's ecological footprint, and Farrell provides an inspiring call to arms, highlighting the ability of students to effect positive change.

We share one Earth—and one future.

Mark White and Phoebe Crisman
Charlottesville, Virginia
December 2013

# Section I

# Introduction and History

# THE PROBLEM OF PRODUCTION

By E. F. Schumacher

> By and large, our present problem is one of attitudes and implements. We are remodeling the Alhambra with a steam-shovel, and are proud of our yardage. We shall hardly relinquish the shovel, which after all has many good points, but we are in need of gentler and more objective criteria for its successful use.
>
> —Aldo Leopold, *A Sand County Almanac*

One of the most fateful errors of our age is the belief that 'the problem of production' has been solved. Not only is this belief firmly held by people remote from production and therefore professionally unacquainted with the facts—it is held by virtually all the experts, the captains of industry, the economic managers in the governments of the world, the academic and not-so-academic economists, not to mention the economic journalists. They may disagree on many things but they all agree that the problem of production has been solved; that mankind has at last come of age. For the rich countries, they say, the most important task now is 'education for leisure' and, for the poor countries, the 'transfer of technology'.

That things are not going as well as they ought to be going must be due to human wickedness. We must therefore construct a political system so perfect that human wickedness disappears and everybody behaves well, no matter how much wickedness there may be in him or her. In fact, it is widely held that everybody is born good; if one turns into a criminal or an exploiter, this is the fault of 'the system'. No doubt 'the system' is in many ways bad and must be changed. One of the main reasons why it is bad and why it can still survive in spite of its badness, is this erroneous view that the 'problem of production' has been solved. As this

3

error pervades all present-day systems there is at present not much to choose between them.

The arising of this error, so egregious and so firmly rooted is closely connected with the philosophical, not to say religious, changes during the last three or four centuries in man's attitude to nature. I should perhaps say: western man's attitude to nature, but since the whole world is now in a process of westernisation, the more generalised statement appears to be justified. Modern man does not experience himself as a part of nature but as an outside force destined to dominate and conquer it. He even talks of a battle with nature, forgetting that, if he won the battle, he would find himself on the losing side. Until quite recently, the battle seemed to go well enough to give him the illusion of unlimited powers, but not so well as to bring the possibility of total victory into view. This has now come into view, and many people, albeit only a minority, are beginning to realise what this means for the continued existence of humanity.

The illusion of unlimited power, nourished by astonishing scientific and technological achievements, has produced the concurrent illusion of having solved the problem of production. The latter illusion is based on the failure to distinguish between income and capital where this distinction matters most. Every economist and businessman is familiar with the distinction, and applies it conscientiously and with considerable subtlety to all economic affairs—except where it really matters—namely, the irreplaceable capital which man had not made, but simply found, and without which he can do nothing.

A businessman would not consider a firm to have solved its problems of production and to have achieved viability if he saw that it was rapidly consuming its capital. How, then, could we overlook this vital fact when it comes to that very big firm, the economy of Spaceship Earth and, in particular, the economies of its rich passengers?

One reason for overlooking this vital fact is that we are estranged from reality and inclined to treat as valueless everything that we have not made ourselves. Even the great Dr. Marx fell into this devastating error when he formulated the so-called 'labour theory of value'. Now, we have indeed laboured to make some of the capital which today helps us to produce—a large fund of scientific, technological, and other knowledge; an elaborate physical infrastructure; innumerable types of sophisticated capital equipment, etc.—but all this is but a small part of the total capital we are using. Far larger is the capital provided by nature and not by man—and we do not even recognise it as such. This larger part is now being used up at an alarming rate, and that is why it is an absurd and suicidal error to believe, and act on the belief, that the problem of production has been solved.

Let us take a closer look at this 'natural capital'. First of all, and most obviously, there are the fossil fuels. No one, I am sure, will deny that we are treating them as income items although they are undeniably capital items. If we treated them as capital items, we should be concerned with conservation: we should do everything in our power to try and minimise their current rate of use; we might be saying, for instance, that the money obtained from the realisation of these assets—these irreplaceable assets—must be placed into a special fund to be devoted exclusively to the evolution of production methods and patterns of living which do not depend on fossil fuels at all or depend on them only to a very slight extent. These and many other things we should be doing if we treated fossil fuels as capital and not as income. And we do not do any of them, but the exact contrary of every one of them: we are not in the least concerned with conservation: we are maximising, instead of minimising the current rates of else; and, far from being interested in studying the possibilities of alternative methods of production and patterns of living—so as to get off the collision course on which we are moving with ever-increasing speed—we happily talk of unlimited progress along the beaten track of 'education for leisure' in the rich countries, and of 'the transfer of technology' to the poor countries.

The liquidation of these capital assets is proceeding so rapidly that even in the allegedly richest country in the world, the United States of America, there are many worried men, right up to the White House, calling for the massive conversion of coal into oh and gas, demanding ever more gigantic efforts to search for and exploit the remaining treasures of the earth. Look at the figures that are being put forward under the heading 'World Fuel Requirements in the Year 2000'. If we are now using something like 7,000 million tons of coal equivalent, the need in twenty-eight years' time will be three times as large—around 20,000 million tons! What are twenty-eight years? Looking back-wards, they take us roughly to the end of World War II, and, of course, since then fuel consump-tion has trebled; but the trebling involved an increase of less than 5,000 million tons of coal equivalent. Now we are calmly talking about an increase three times as large.

People ask: can it be done? And the answer comes back: it must be done and therefore it shall be done. One might say (with apologies to John Kenneth Galbraith) that it is a case of the bland leading the blind. But why cast aspersions? The question itself is wrong-headed, because it carries the implicit assumption that we are deal-ing with income and not with capital. What is so special about the year 2000? What about the year 2028, when little children running about today will be planning for their retirement? Another trebling by then? All these questions and answers are seen to be absurd the moment we realise that we are dealing with capital and not with income: fossil fuels are not made by men; they cannot be recycled. Once they are

gone they are gone forever. But what—it will be asked—about the income fuels? Yes, indeed, what about them? Currently, they contribute (reckoned in calories) less than four per cent to the world total. In the foreseeable future they will have to contribute seventy, eighty, ninety per cent. To do something on a small scale is one thing: to do it on a gigantic scale is quite another, and to make an impact on the world fuel problem, contributions have to be truly gigantic. Who will say that the problem of production has been solved when it comes to income fuels required on a truly gigantic scale?

Fossil fuels are merely a part of the 'natural capital' which we steadfastly insist on treating as expendable, as if it were income, and by no means the most' important part. If we squander our fossil fuels, we threaten civilisation; but if we squander the capital represented by living nature around us, we threaten life itself. People are waking up to this threat, and they demand that pollution must stop. They think of pollution as a rather nasty habit indulged in by careless or greedy people who, as it were, throw their rub-bish over the fence into the neighbour's garden. A more civilised behaviour, they realise, would incur some extra cost, and therefore we need a faster rate of economic growth to be able to pay for it. From now on, they say, we should use at least some of the fruits of our ever-increasing productivity to improve 'the quality of life' and not merely to increase the quantity of con-sumption. All this is fair enough, but it touches only the outer fringe of the problem

To get to the crux of the matter, we do well to ask why it is that all these terms—pollution, environment, ecology etc—have so suddenly come into prominence. After all, we have had an industrial system for quite some time, yet only five or ten years ago these words were virtually unknown. Is this a sudden fad, a silly fashion, or perhaps a sudden failure of nerve?

The explanation is not difficult to find. As with fossil fuels, we have indeed been living on the capital of living nature for some time, but at a fairly modest rate. It is only since the end of World War II that we have succeeded in increasing this rate to alarming proportions. In comparison with what is going on now and what has been going on progressively, during the last quarter of a century, all the industrial activities of mankind up to, and including, World War II are as nothing. The next four or five years are likely to see more industrial production, taking the world as a whole, than all of mankind accomplished up to 1945. In other words, quite recently that most of us have hardly yet become conscious of it—there has been a unique quantitative jump in industrial production.

Partly as a cause and also as an effect, there has also been a unique qualitative jump. Our scientists and technologists have learned to compound substances unknown to nature; against many of them, nature is virtually defenceless. There are no natural agents to attack and break them down. It is as if aborigines were suddenly attacked with machine-gun fire: their bows and arrows are of no avail. These substances, unknown to nature, owe their almost magical effectiveness precisely to nature's defencelessness—and that accounts also for their dangerous ecological impact. It is only in the last twenty years or so that they have made their appearance in bulk. Because they have no natural enemies, they tend to accumulate, and the long-term consequences of this accumulation are in many cases known to be extremely dangerous, and in other Gases totally unpredictable.

In other words, the changes of the last twenty-five years, both in the quantity and in the quality of man's industrial processes, have produced an entirely new situation—a situation resulting not from our failures but from what we thought were our greatest successes.

And this has come so suddenly that we hardly noticed the fact that we were very rapidly using up a certain kind of irreplaceable capital asset, namely the tolerance margins which benign nature always provides.

Now let me return to the question of 'income fuels' with which I had previously dealt in a somewhat cavalier manner. No one is suggesting that the world-wide industrial system which is being envisaged to operate in the year 2000, a generation ahead, would be sustained primarily by water or wind power. No, we are told that we are moving rapidly into the nuclear age. Of course, this has been the story for quite some time, for over twenty years, and yet. the contribution of nuclear energy to man's total fuel and energy requirements is still minute. In 1970, it amounted to 27 per cent in Britain; 0·6 per cent in the European Community; and 0·3 per cent in the United States, to mention only the countries that have gone the furthest. Perhaps we can assume that nature's tolerance margins will be able to cope with such small impositions, although there are many people even today who are deeply worried, and Dr. Edward D. David, President Nixon's Science Adviser, talking about the storage of radioactive wastes, says that 'one has a queasy feeling about something that has to stay underground and be pretty well sealed off for 25,000 years before it is harmless'.

However that may be, the point I am making is a very simple one: the proposition to replace thousands of millions of tons of fossil fuels, every year, by nuclear energy means to 'solve' the fuel problem by creating an environmental and ecological problem of such a monstrous magnitude that Dr. David will not be the only one to have 'a queasy feeling'. It means solving one problem by shifting it to another sphere—there to create an infinitely bigger problem.

Having said this, I am sure that I shall be confronted with another, even more daring proposition: namely, that future scientists and technologists will be able to devise safety rules and precautions of such perfection that the using, transporting, processing and storing of radioactive materials in ever-increasing quantities will be made entirely safe; also that it will be the task of politicians and social scientists to create a world society in which wars or civil disturbances can never happen. Again, it is a proposition to solve one problem simply by shifting it to another sphere, the sphere of everyday human behaviour. And this takes us to the third category of 'natural capital' which we are recklessly squandering because we treat it as if it were income: as if it were something we had made ourselves and could easily replace out of our much-vaunted and rapidly rising productivity.

Is it not evident that our current methods of production are already eating into the very substance of industrial man? To many people this is not at all evident. Now that we have solved the problem of production, they say, have we ever had it so good? Are we not better fed, better clothed, and better housed than ever before—and better educated! Of course we are: most, but by no means all of us: in the rich countries. But this is not what I mean by 'substance'. The substance of man cannot be measured by Gross National Product. Perhaps it cannot be measured at all except for certain symptoms of loss. However, this is not the place to go into the statistics of these symptoms, such as crime, drug addiction, vandalism, mental breakdown, rebellion, and so forth. Statistics never prove anything.

I started by saying that one of the most fateful errors of our age is the belief that the problem of production has been solved. This illusion, I suggested, is mainly due to our inability to recognise that the modern industrial system, with all its intellectual sophistication, consumes the very basis on which it has been erected. To use the language of the economist, it lives on irreplaceable capital which it cheerfully treats as income. I specified three categories of such capital: fossil fuels, the tolerance margins of nature, and the human substance. Even if some readers should refuse to accept all three parts of my argument, I suggest that any one of them suffices to make my case.

And what is my case? Simply that our most important task is to get off our present collision course. And who is there to tackle such a task? I think every one of us, whether old or young, powerful or powerless, rich or poor, influential or uninfluential. To talk about the future is useful only if it leads to action *now*. And what can we do *now*, while we are still in the position of 'never having had it so good'? To say the least—which is already very much—we must thoroughly understand the problem and begin to see the possibility of evolving a new life-style, with new methods of production and new patterns of consumption: a life-style designed for permanence. To give only three preliminary examples: in agriculture and horticulture, we can interest ourselves in the perfection of production methods which are biologically sound, build up soil fertility, and produce health, beauty and permanence. Productivity will then look after itself. In industry, we can interest ourselves in the evolution of small-scale technology, relatively non-violent technology, 'technology with a human face', so that people have a chance to enjoy themselves while they art: working, instead of working solely for their pay packet and hoping, usually forlornly, for enjoyment solely during their leisure time. In industry, again—and, surely, industry is the pace-setter of modern life—we can interest ourselves in new forms of partnership between management and men, even forms of common ownership.

We often hear it said that we are entering the era of 'the Learning Society'. Let us hope this is true. We still have to learn how to live peacefully, not only with our fellow men but also with nature and, above all, with those Higher Powers which have made nature and have made us; for, assuredly, we have not come about by accident and certainly have not made ourselves.

The themes which have been merely touched upon in this chapter will have to be further elaborated as we go along. Few people will be easily convinced that the challenge to man's future cannot be met by making marginal adjustments here or there, or, possibly, by changing the political system.

The following chapter is an attempt to look at the whole situation again, from the angle of peace and permanence. Now that man has acquired the physical means of self-obliteration, the question of peace obviously looms larger than ever before in human history. And how could peace be built without some assurance of permanence with regard to our economic life?

# FROM MUIR TO MEADOWS

By Simon Dresner

I think that God in creating man somewhat overestimated his ability.
— Oscar Wilde

## John Muir and the Sierra Club

The organized environmental movement was started by John Muir, a
Scottish emigrant to America. In
1864, as a young man of 26, he
disappeared into the Great Lakes wilderness.
There, he discovered an awe for nature similar to
that of the Romantics earlier. Muir later walked
across much of North America, eventually going
to California, where he settled and focused his
energy on preserving the Sierras.

Muir's first book, *The Mountains* of
*California,* describes the natural wonder of the
state and also the loss of biological diversity in
California already apparent due to the pressure
of development:

But of late years plows and sheep
have made sad havoc in these
glorious pastures, destroying tens of
thousands of the flowery acres like
a fire, and banishing many species of
the best honey-plants to rocky-cliffs
and fencecorners, while, on the other
hand, cultivation thus far has given no
adequate compensation, at least in
kind; only acres of alfalfa for miles of
the richest wild pasture, ornamental
roses and honeysuckles around
cottage doors for cascades of wild
roses in the dells, and small, square
orchards and orange groves for broad
mountain-belts of chaparral.[1]

Muir's greatest personal achievement against
the forces of development was the establish-
ment of Yosemite as a National Park in 1890.
Within two years development pressures were
so severe that a group of Californians led by

Muir founded the Sierra Club to defend the park. The final years of Muir's life were dominated by the unsuccessful 1908–13 campaign to prevent the building of a dam in Hetch Hetchy Valley, the next valley north of Yosemite.

## Gifford Pinchot's Conservationism

A very different tradition of conservation also appeared in the United States at the turn of the century, personified by Gifford Pinchot. He was the most prominent advocate of 'sustained yield' forestry, in contrast to the rapacious practices of the 'robber barons'. The Conservationist approach he put forward was implemented under the presidency of Teddy Roosevelt in the first decade of the 20th century. Roosevelt appointed Pinchot the first director of the US Forest Service. His Conservationism sought to conserve natural resources as it saw the destruction of resources such as forests as wasteful. The justification for conservation was that the resources could be more economically efficiently exploited. Pinchot said that the aim of conservation, echoing Jeremy Bentham's formulation of utilitarianism, was 'the greatest good for the greatest number for the longest time'. The Conservationist position was known as 'wise use' and became the cornerstone of official thinking about the environment in the United States. It was criticized even from the start by more radical thinkers, often called Preservationists, led by John Muir. The Conservationists believed that not utilizing natural resources was wasteful. When it was suggested that hydroelectric development of Niagara would spoil the river's beauty, Conservationists responded that it would be a crime to let so much energy go to waste.[2] The two groups fought over the Hetch Hetchy dam, intended to provide water for the rapidly growing city of San Francisco. It was a battle that eventually the Preservationists lost.

The Conservationists had supported the dam, claiming it was a wise use of nature for human ends. The Preservationists argued against its construction as it would destroy one of the most beautiful valleys in the world. John Muir personally believed that the crime in building the dam was the destruction of nature more than the loss of aesthetic value. At that time, though, the use of this non-anthropocentric argument was not politically viable.

The Conservationists invented the profession of forestry, which they saw as tree farming. They created America's National Forests, areas of public land set aside primarily to provide for the need for timber. To obtain the maximum 'sustained yield' of timber, foresters grow only one species of tree in neat rows to make for easier harvesting and operate a system of rotation every few decades. The result lacks all the biological diversity of natural forests.

The practical problems with Conservationism lay in its lack of understanding of ecology. The Conservationists brought about the adoption by the US government of the policy of exterminating all predators. It became apparent to some people that this was not necessarily a good idea when it led to the population of deer in the Kaibab Forest near the Grand Canyon exploding, then crashing after their food supply was destroyed in the 1920s.[3] One response to these problems was the development of ecology as a science.

## Aldo Leopold's Land Ethic

Aldo Leopold decided that the problem lay deeper than insufficient scientific knowledge. Leopold had been trained by Conservationists in the Yale Forestry School, and for the early part of his professional life he accepted Conservationist teaching. Gradually, however, he became disillusioned. The issue that precipitated this change of heart was predator control.[4] In his book *A Sand County Almanac,* Leopold called for a 'land ethic'.[5] Leopold's intuition was that the earth was an indivisible living being, each species playing its part in an indivisible whole, and human beings were just one part of that community. The land ethic stated that 'a thing is right when it tends to preserve the integrity, stability, and beauty of the biotic community. It is wrong when it tends otherwise.'[6] He said that the land ethic 'changes the role of *Homō sapiens* from conqueror of the land-community to plain member and citizen of it'.[7]

Yet Leopold justified the land ethic in anthropocentric terms—essentially that the long-term human interest was best served by a healthy ecosystem, even if the short-term interest was best served by purely economic criteria. That is the argument given for adopting criteria of 'sustainability' in decision making today. Leopold argued that it was not sensible from our point of view to remove a species from an ecosystem without knowing what the long-term consequences would be: 'To keep every cog and wheel is the first precaution of intelligent tinkering.'[8] That is another argument that is familiar today, in the form of the precautionary principle.

## RACHEL CARSON'S *SILENT SPRING*

The new environmental movement that emerged in the 1960s was sparked off by Rachel Carson and her book *Silent Spring.*[9] Carson was a respected writer and scientist who wrote the book to draw attention to the destruction of wildlife by the use of the pesticide DDT. What was new about Carson's book was that it criticized a technology intended to better the condition of the human race, rather than a specific development, and that her book revealed

*unintended and unpredicted* consequences of this technology.

Carson revealed that our actions could lead to seriously damaging environmental consequences when we interfered with natural systems we did not fully understand. She criticized the unthinking use of the technological 'quick fix' of employing synthetic chemicals to control insects. She warned that these chemicals contained the prospect of a dying world in which springtime would no longer bring forth new life, only silence. Carson concluded that 'the "control of nature" is a phrase conceived in arrogance, born of the Neanderthal age of biology and philosophy, when it was supposed that nature exists for the convenience of man'.[10] Carson's challenge to pesticides was implicitly a challenge to science and the idea of technological progress.

Modern environmentalism has two key concerns: the limits to control that were emphasized by Leopold and Carson, and also the idea of a global environmental crisis—limits of scale on a small planet. Atmospheric nuclear testing was banned in 1963 after the discovery of strontium-90 from nuclear fallout in mother's milk and in the fat of Antarctic penguins. Not long afterwards, traces of DDT were also found in Antarctic penguins. The notion that the world was not large, but relatively small, began to gain currency.

## SPACESHIP EARTH

The metaphor of the earth as a spaceship was coined by the American presidential candidate Adlai Stevenson in a campaign speech as far back as 1952.[11] It was taken up by the British journalist Barbara Ward in her 1966 book *Spaceship Earth.*[12] The concept was developed simultaneously by the economist Kenneth Boulding. In his essay 'The Economics of the Coming Spaceship Earth' Boulding put forward the idea that previous human history had taken place against a background where the scale of human activities was tiny compared to the environment.[13] In this situation, there was always somewhere 'out there' to expand to or put your wastes. He called this a 'cowboy' economy because the idea of an endless frontier was embodied by the American cowboy. What was now happening was that human activity was growing to a size where there was no 'out there' left. In this situation it was no longer possible to try to put problems somewhere else. They would always return to you. He wrote that it would require a 'spaceman' economy because frontiers had shrunk to zero, there was nowhere to expand to, nothing could be simply thrown away and all waste would have to be recycled. Boulding's essay was influential, but the metaphor did not catch on immediately.

In the 1960s there was a growing sense that Western technology had reached every corner of the earth and that with improved communication the world was growing smaller and closer together. This was the time of Marshall McLuhan's 'global village'.[14] It was also the time when a global nuclear holocaust seemed a real possibility. What really brought the metaphor alive was the photographs of the Earth that the Apollo astronauts took from the Moon at the end of the 1960s. The pictures were of a small and beautiful blue-white planet with oceans and clouds, against the blackness of space and the grey of another, lifeless, world. The photograph perfectly visualized the metaphor 'Spaceship Earth' and the environmental movement seized on it. Within a few years the metaphor had become a tiresome cliché, but while it lasted it was extremely powerful. In 1970 the first Earth Day was held. In 1972, at the UN Conference on the Human Environment, the official slogan was 'Only One Earth'.

The metaphor has two different connotations. One is of limits to human activities. The other is of the need for human management of the environment. The two meanings are not entirely incompatible, but there is clearly a tension between them. One implies that we are overcrowding passengers. The other implies that we are the new commander about to bravely go where no species has gone before, presumably replacing God.

## THE COSTS OF ECONOMIC GROWTH

In 1967, the prominent economist Edward Mishan shook up the economics profession with the publication of *The Costs of Economic Growth*.[15] He argued that calculations of Gross National Product (GNP) were seriously misleading as a measure of human welfare because they included the costs of defensive measures such as anti-pollution expenditure and failed to count negative effects of affluence like aircraft noise against growth. A decade earlier, John Kenneth Galbraith had made fairly similar arguments against GNP as a measure of welfare, but without the emphasis on environmental externalities.[16] Mishan's argument was sound, but embarrassing to economists.

## THE TRAGEDY OF THE COMMONS

The ecologist Garrett Hardin's 1968 essay discussed what he called 'the tragedy of the commons'.[17] He described a hypothetical example of a pasture shared by herders. It is in the interests of each herder to put additional animals out to pasture because they will get additional income from them, However, each individual deciding to do that will lead to the pasture being overgrazed and degraded, but there is no incentive for them not to because others will do it if they don't. The tragedy of the commons can only be solved if access can somehow be controlled. Hardin stated that the tragedy of the commons was one example of many of the problems created by human population growth and increasing consumption of natural resources.

## The Population Bomb

The era of 'Spaceship Earth' was the time when fear of global environmental limits began to emerge. Environmentalism came together with a renewed Malthusianism. Concern about exponential population growth, this time in the developing world, came to the fore with the publication of Paul Ehrlich's *The Population Bomb*.[18] Paul and Anne Ehrlich (who co-wrote the book but was denied credit by the publisher) were animal population biologists who had developed these opinions after personally witnessing the hunger and overcrowding in India. The Ehrlichs argued that population growth would lead to massive famines in Asia and Africa in the 1970s. Most controversially, they proposed that some countries, such as India, were such hopeless cases that resources should not be wasted on helping them. This ruthless position was very reminiscent of Malthus himself, the first person to suggest that exponential growth would lead to an imminent collision with natural limits.

The Ehrlichs' terrible predictions of famine and imminent ecological collapse failed to come true. Famines had occurred quite often in India until the 1970s, but have not happened since (although the Ehrlichs point out that the present population of India is being fed by methods that cause massive soil erosion[19]). The large famines of the 1970s and 1980s in Africa took place mostly in war-torn countries. It seems, though, that at the time the self-proclaimed alarmism of *The Population Bomb* was very influential in raising concern. The theologian and environmentalist John Cobb told me that it was *The Population Bomb's* alarmism that motivated him to become an environmental activist. Cobb said that although the book had been wrong in its predictions and contained 'gross exaggerations', he wanted to acknowledge that it was *The Population Bomb* that shook him into action. The environmental economist John Pezzey was also scared into environmentalism by *The Population Bomb* and other alarmist books of the late 1960s. He told me that he now felt cheated by the alarms that were raised about problems that had not turned out so bad as he had been led to believe then. This kind of alarmism was an easy target for John Maddox, the editor of the journal *Nature,* in his anti-environmentalist polemic *The Doomsday Syndrome*.[20] Today, his optimism appears as wide of the mark as the Ehrlichs' pessimism. Maddox did not believe that there were ecological limits. For example, he argued that there was no shortage of land for cultivation in the world and pointed to the Amazon and Congo basins as large areas suitable for agricultural development.

Ehrlich went on to develop his ideas to consider the overall environmental impact of the human population, rather than sheer numbers. It was expressed in the equation I=PCT, where I is the environmental impact, P is the population size, C is per capita consumption and T is the environmental impact of the productive technology.[21] Although many of Ehrlich's early ideas are regarded with embarrassment by most environmentalists today, the I=PCT equation opened up the way to many of the ideas that environmentalists campaign on today because the levels of material consumption in Western countries are several times the global average and 50 times those in the poorest countries. I=PCT actually suggests that increasing consumption is a greater cause of the growing environmental impact of the human race than population growth.

## THE LIMITS TO GROWTH

The crystallization of the concerns of the first wave of environmentalism that ran from around 1966 to 1972 was The Limits to Growth.[22] A report by a group of young scientists from the Massachusetts Institute of Technology (MIT), it immediately took the world by storm, gaining enormous media coverage. It was translated into 28 languages and sold 9 million copies. The Limits to Growth was based on computer models that appeared to show that if the current trends of exponential growth in population and demand for non-renewable resources continued, the world would face severe shortages of food and non-renewable resources by the middle of the 21st century. The modellers concluded that 'the limits to growth on this planet will be reached some time in the next hundred years'.[23]

The model assumed that there are finite amounts of fossil fuels and minerals available on the Earth. It did not simply assume that you use a certain amount and then run out, but modelled the price behaviour of an increasingly scarce and difficult to obtain resource. The estimates of availability given could be challenged but, the MIT team argued, because of the nature of exponential growth, even if resources were several times larger than the current estimates, they would still become extremely scarce and expensive only a few decades later. According to the model, population growth was happening too fast for demographic transition before collapse unless population control measures were introduced. The authors admitted they had no idea of the pollution absorption capacity of the environment, but they felt that with exponential growth in pollutants it would be reached relatively quickly. They modelled pollution as a single long-lived chemical that in high concentrations would shorten human life and interfere with food production.

The authors explicitly stated that their model was not a definite prediction of what would happen—it was an exploration of the consequences of current trends. They ran versions of the model which assumed various changes, such as enormous potential increases in agricultural productivity, the availability of cheap nuclear power, extensive mineral recycling and very strict pollution control standards. Even with all these running in the model, exponential growth still caused an overshoot of what could be sustainably supported by the planet and a collapse of civilization before the end of the 21st century. When they modelled a future with zero population growth and zero capital growth, and assuming a four-fold increase in the technological efficiency of production, and investment in agriculture to end malnutrition, the model gave an ultimately stable state at a European average standard of living: 'It is possible to alter these growth trends and to establish a condition of ecological and economic stability that is sustainable far into the future.'[24]

## CRITICISM OF THE LIMITS TO GROWTH

Critics pointed out that the output of the computer model was determined by the assumptions the programmers had made. The project had been funded by the Club of Rome, an international grouping of prominent scientists, business people and civil servants concerned about environmental problems. They were of an essentially Malthusian persuasion. Donella Meadows stated that she and her husband were Malthusians when they started the project. In fact, they started work on the project two weeks after returning from a year in Asia. Donella Meadows told me about the conclusions of the model:

> We had kind of intuited it in India, but it was just a feeling to go from one of the world's richest places to one of the world's poorest and to see the soil erosion and to see the children, the burgeoning cities and the disappearing forests. You just somehow knew all of this was inconsistent, was offensive, was morally intolerable and furthermore it was physically impossible to continue doing things this way. We knew that, but we couldn't put a case for it. It was something any intelligent observer knows anywhere they go in the world. So the computer model helped us to put numbers, to put time frames, to get a much neater mental model of the problem and the possible solutions.

It can be argued that they already knew what they were expecting to find and wrote the model in that way. There is perhaps truth in that. Because of the status computers had at that time, there was a tendency for the public to believe any computer model was correct. The reality was that it all depended on the validity of the assumptions.

The best developed critique of *The Limits to Growth* came from a team at Sussex University's Science Policy Research Unit.[25] They pointed out that the computer model was no less subjective or ideological than the mental models on which it was based. They criticized several aspects of the model, but concentrated on the Malthusian pessimism of the assumptions underlying it. The Sussex team examined the model and argued that the assumptions about the rate of technological progress and the availability of physical resources were too

pessimistic. They accepted that physical growth cannot continue indefinitely on a finite planet, but held that any physical limits were much more distant. The Sussex team more generally criticized the determinism of the model as it did not include the feedbacks that would allow for resource substitution, new inventions and changing ways of life. They accused *The Limits to Growth* of discounting the potential for adaptation in human society and putting forward a counsel of despair in proposing an immediate end to growth. The Sussex team claimed that the concentration on disaster in 100 years' time distracted from what could be done to solve urgent existing problems, such as the distribution of the world's wealth.

## HERMAN DALY'S *STEADY-STATE ECONOMICS*

The ideas in *The Limits to Growth* did not go away, though. They were instead taken up and elaborated. Herman Daly used the law of entropy to attempt to demonstrate that the scale of the economy was limited.[26] Economic activity is about the creation of order (low entropy) in one place. The entropy law demands that a larger amount of entropy is created elsewhere. Daly's former professor Nicholas Georgescu-Roegen had already used that argument.[27] He had concentrated on the irreversibility of the use of non-renewable fossil fuels as a source of energy. Daly made the point that economic activity (or rather energy and material throughput) necessarily creates pollution and wastes. More activity means more pollution and waste. There is a limit to how much the biosphere can absorb. Daly concluded that the entropy law set a limit to the scale of the economy. His claim that entropy sets a limit to the physical scale of the economy is now widely accepted, but his conclusion that there is an absolute limit to economic growth is still very controversial.

## GLOBAL 2000

Ideas about limits to growth even influenced the 1977–81 Carter administration in the United States. President Carter was concerned about the 'energy crisis' and promoted research into renewable energy sources. He commissioned a report on the state of the global environment up to 2000. The report's conclusion was:

> If present trends continue, the world in 2000 will be more crowded, more polluted, less stable ecologically and more vulnerable to disruption. Serious stresses involving population, resources, and environment are clearly visible ahead. Despite greater material output, the world's people will be poorer in many ways than they are today.

> For hundreds of millions of the desperately poor, the outlook for food and other necessities of life will be no better. For many it will be worse. Barring revolutionary advances in technology, life for most people on earth will be more precarious in 2000 than it is now—unless the nations of the world act decisively to alter current trends. [28]

The finding from the *Global 2000* study which came as a surprise to many people was a calculation that habitat destruction was likely to lead to the extinction of 500,000 to 2 million species, mostly in tropical forests. Concern about loss of biodiversity, first raised by the biologist Norman Myers, quickly moved up to become a major environmental concern. [29]

At the end of 1980, Carter lost power to Ronald Reagan, who was a determined anti-environmentalist. Campaigning for the Presidency, he claimed: 'Approximately 80 per cent of our air pollution stems from hydrocarbons released by vegetation. So let's not go overboard in setting and enforcing tough emissions standards for man-made sources.' Once elected President, he went on to say: 'Trees cause more pollution than automobiles do.'[30] These beliefs were reflected in the policies that his administration followed.

As a consequence, environmental leadership passed from the United States to Europe in the 1980s. The environmental movement had already spread in the early 1970s, first to Scandinavia, then to the rest of Western Europe, and particularly to Germany. European environmentalism was less concerned with the wilderness issues that have always remained crucial to North American environmentalism, and was more concerned with the problems of industrialism.

In the 1970s, though, environmentalism was perceived as a Western idea of little interest to the developing world. Environmentalists invented the concept of sustainability in an attempt to overcome hostility to their concerns in the developing world.

## NOTES

1. John Muir (1894) *The Mountains of California*, San Francisco: Sierra Club, 1988, p 257

2. Juan Martinez-Alier (1987) *Ecological Economics*, Oxford: Basil Blackwell

3. Donald Worster (1985) *Nature's Economy*, Cambridge: Cambridge University Press

4. Roderick Nash (1989) *The Rights of Nature*, Madison: University of Wisconsin Press

5. Aldo Leopold (1949) *A Sand County Almanac*, New York: Ballantine, 1970

6. ibid, p 262

7. ibid, p 240

8. ibid, p 190

9. Rachel Carson (1962) *Silent Spring*, New York: Houghton Mifflin

10. ibid, p 243

11. Richard B. Norgaard (1994) *Development Betrayed*, London: Routledge

12. Barbara Ward (1966) *Spaceship Earth*, New York; University of Columbia Press

13. Kenneth Boulding (1966) 'The Economics of the Coming Spaceship Earth' in *Environmental Quality in a Growing Economy*, edited by H. Jarrett, Baltimore: Johns Hopkins Press

14. Marshall McLuhan and Quentin Fiore (1967) *The Medium is the Massage*, New York: Bantam

15. E. J. Mishan (1967) *The Costs of Economic Growth*, London: Staples

16. John Kenneth Galbraith (1958) *The Affluent Society*, New York: New American Library

17. Garrett Hardin (1968) 'The Tragedy of the Commons', *Science*, 162: 1243–1248

18. Paul Ehrlich (1968) *The Population Bomb*, New York: Ballantine

19. Paul R. Ehrlich and Anne H. Ehrlich (1991) *The Population Explosion*, New York: Simon & Schuster

20. John Maddox (1972) *The Doomsday Syndrome*, London: Macmillan

21. John Holdren and Paul Ehrlich (1974) 'Human population and the global environment', *American Scientist*, (62) 282–292

22. Donella H. Meadows, Dennis L. Meadows, Jørgen Randers and William W. Behrens III (1972) *The Limits to Growth*, New York: Universe Books

23. ibid, p 23

24. ibid, p 24

25. H. S. D. Cole, Christopher Freeman, Marie Jahoda and K. L. R. Pavitt (1973) *Thinking About The Future: A Critique of 'The Limits of Growth'*, London: Chatto & Windus

26. Herman E. Daly (1977) *Steady-State Economics*, San Francisco: Freeman

27. Nicholas Georgescu-Roegen (1971) *The Entropy Law and the Economic Process*, Cambridge, Massachusetts: Harvard University Press

28. Gerald Barney, director (1981) *Global 2000 Report to the President*, New York: Penguin, 1

29. Norman Myers (1979) *The Sinking Ark*, New York; Pergamon

30. Mark Green and Gail MacColl (1983) *There He Goes Again: Ronald Reagan's Reign of Error*, New York: Pantheon

# POPULATION, CONSUMPTION, AND THE ENVIRONMENT

By Paul K. Conkin

Almost any consideration of the earth's present health, or its prospects during the next century, has to begin with the human population. The doubling of the world's population between 1960 and 2000, the 6.5 billion people on earth in 2006, and the prospect of 9 billion by 2050 raise innumerable issues about available resources, about the level of pollution and waste, about massive extinctions, and about the quality of human life in crowded cities. Countries with nearly stable or even declining populations do not face some of these problems, but these are the very countries with the highest levels of consumption, resource use, and emissions. They also have economies that are predicated on a continued growth in living standards. The pressures on the earth thus come from both directions, from the multiplying poor and the indulgent rich.

## POPULATION AND RESOURCES

It is much too early to assess with any degree of assurance the consequences of the present population explosion. Such a new surge of population growth is not new, but its pace has been unprecedented. The first surge may have begun even before the evolution of Homo sapiens, when humanoids first learned to control fire. The second surge in population began when humans moved from hunting and gathering to the domestication of animals and to the cultivation of crops. The present surge was only the climax of a more gradual expansion of population in the modern era, particularly in the nineteenth century. What changed is that after 1950 so many trend lines turned sharply upward. One example is what happened to agriculture in the developed countries, and particularly in the United States, where productivity almost doubled from 1950 to 1970 because of increased uses of chemicals for fertilizer and pest control, new and more productive varieties of crops, and the use of fossil fuels to power larger and more efficient machines. The green revolution spread. Since 1970 the world's production of food has more than doubled. Without this agricultural revolution, the earth simply could not feed the present population, and in a sense is not even feeding it well in the present (over 800 million people are hungry because they have to survive at less than an optimum level of nutrition).

After past introductions of new technologies, the subsequent growth of population soon leveled off. In effect, larger populations eventually probed the existing limits of subsistence. Will the present population explosion soon level off? Obviously, the growth rate cannot continue at present levels, and is already slowing in most countries, with worldwide annual growth rates down from 2.1 percent in 1970 to 1.14 percent in 2004 (see figures 1 and 2). Even the 9 billion expected by 2050 reflects more than a 50 percent decrease in the rate of growth from that

of the last half of the twentieth century. In 2003, in the wake of the AIDS epidemic, the United Nations Population Division lowered its median estimate for 2050 from 9.3 to 8.9 billion. But in its 2004 revision, its medium projected 2050 population is back up to 9.1 billion. Even 9 billion people will present new problems. In fact, two-thirds of the world's population is already pressing against such intractable resource scarcities, and such environmental degradation, as to make even low incomes difficult to increase. In the poorest countries, the growth of population has pushed beyond the limits of economic growth, with a bleak future for such populations in the future. At least eighteen countries, most in central Africa, have suffered a negative per capita income growth in the last decade, and at least thirty other countries have enjoyed little if any growth, with an AIDS pandemic aggravating already desperate economic challenges.

One may object that the present problems in poor countries result not primarily from population growth, but from low productivity

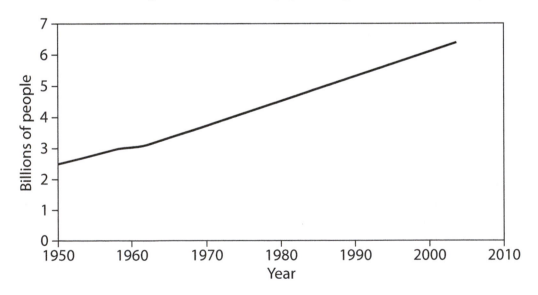

FIGURE 1. WORLD POPULATION, 1950–2004.

Data from Census Bureau. Worldwatch, *Vital Signs* 2005, 65.

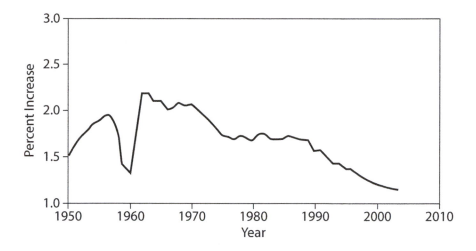

FIGURE 2. ANNUAL GROWTH RATE IN WORLD POPULATION, 1950–2004.

Data from Census Bureau. Worldwatch, *Vital Signs* 2005, 65.

based on a variety of problems that are, in principle, correctable—unstable governments, a lack of educational opportunities and thus a shortage of human capital, the suppression of women, a lack of modern technology and thus great economic inefficiency, and a primitive agriculture. Demographers have long predicted, or at least hoped, that the same demographic transition that occurred in Europe and North America—from high mortality and birthrates, to technological changes that led to both prosperity and lowered mortality rates, to a final stage with low mortality and the present low fertility rates—would be duplicated elsewhere. But it may not be possible to duplicate such a transition in much of central Africa and southern Asia By coercion. China has come close to such a transition to low fertility despite low per capita incomes. Elsewhere, lowered fertility rates have accompanied higher incomes, but have dropped only slowly or not at all for very low-income families. The transition has taken

place only among elites. The only likely way to gain something close to re-placement level fertility in much of the world seems to depend on rapid economic growth. But it is simply impossible to imagine the natural re-sources—water, soil, energy—that would allow these crowded countries to gain living standards comparable to those in western Europe, at least short of magical new technologies, such as cheap and plentiful fusion energy. And even if they attained such a level of prosperity, and their population leveled off at no more than a third above present levels, one wonders about the quality of life possible in such a crowded world.

From a worldwide perspective, the earth probably has enough resources to feed 9 billion people, even with present agricultural knowledge and tools. This may be small consolation to poor countries that have no way to meet their own food needs. The earth has enough fossil fuels to last for the next fifty years, even at the present annually increased rate of use. This may be small consolation to the two-thirds of the earth's population that have, so far, consumed a small share of such fuels, but have been direly affected by the global impact, including recent warming, due to the production and consumption in wealthy countries.

With the sole exception of the United States, the population of the twenty-three wealthiest countries, or those with a 2002 per capita Gross National Income (GNI) of over $15,000, or approximately 15 percent of the total, is stable or declining. The U.S. Census Bureau predicts that by 2025 all the net population increase will be in the recently poorer countries, and until then 98 per cent will be in such countries (much of the other 2 percent will be in the United States). By then the 20 percent of the world population now living in what the United Nations designates as developed countries (roughly those with a present per capita GDP of over $10,000) will drop to 15 percent.

This means that present demographic imbalances will increase, with a very high percentage of people over sixty-five years old in affluent countries. a very small base of youth under fifteen, and a shortage of working-age people. In poorer countries, at present, from 40 to 50 percent of the total population is under fifteen years old, with the population over sixty ranging from only 3 to 7 percent. Over 90 percent of their dependent population is under fifteen. The huge bulge of child-bearing women in the near future assures a continued population growth for the next three decades despite declining birthrates. Conversely, in the most affluent countries the population under fifteen is only 14–20 percent, except in the United States (21.8 percent), the population over sixty from 20 to 25 percent, except in the United States (16.1 percent). The only means of correcting such demographic imbalances would be a speedup of the present outsourcing of work to underdeveloped countries and a major migration of working-age people from the underdeveloped countries into the labor-short developed world, a migration that has already had a major influence on the population of the United States.

## SUSTAINABLE DEVELOPMENT

Today the verbal mantra "sustainable development" is a loaded phrase. Everyone supports it, but few define it in exactly the same way or honestly probe its implications. The central idea is an old one, at least among economists. If people are to remain prosperous over a long time, they must develop and preserve their tools of production, or capital. Otherwise, they will soon use up capital and face declining returns and ultimately bankruptcy. Today, environmentalists have extended this understanding of the vital role of human made tools to non human-created goods, such as soil, water, air, and fuels. Economic growth, both in developed countries and poor countries, will be self-defeating if it involves a using up of nonrenewable resources, such as fossil fuels, or a steady draw down of renewable resources, such as forests or soil nutrients.

For nonrenewable resources, with fossil fuels by far the most important, humans cannot avoid a continued draw down, at least in the near future. In this case, the mandate of sustainability requires enough research and development in the present to find renewable replacements for fossil fuels before they are exhausted. Finally, humans must not emit more pollutants than the environment can safely assimilate. Today, in no area of the world are economies even close to meeting these goals. Poor countries, by necessity, are rapidly using up renewable resources, while the wealth of affluent countries depends upon the past and continuing exploitation of the world's dwindling reserve of fossil fuels.

The goals of sustainability are not new. Human concerns about scarce resources, and about environmental degradation, reach back to the dawn of civilization. Prehistoric people at times were unable to adjust to environmental change, such as cycles of drought, or pressed too strongly against scarce resources and suffered famine, population decline, and cultural bankruptcy. Few present environmental concerns are new, except those created by new technologies (such as ozone-depleting chemicals). But because of the population explosion, never before have so many environmental problems been global in their implications (global warming, massive extinctions of species, rain forest destruction, acid rain, ocean pollution) and so difficult both to understand and to mitigate.

Sustained development, to the extent that it means economic growth, poses the most difficult challenge for poorer countries. In 2004, the most wealthy twenty countries, with a per capita Gross National Income (GNI, or what was formerly called GNP) of over $25,000 in current U.S. dollars (excluding tiny nations like Lichtenstein), made up less than 15 per cent of the world's population, but they controlled 72 percent of the world's total income. The United States alone accounted for 30 percent of this income. The list of the twenty most wealthy countries includes the United States and Canada. Australia, fifteen western European countries, and only Japan and Hong Kong in Asia, Ii includes no countries in Africa or Latin America, and no country from the former Soviet bloc. In fact, none of these areas have any countries among the additional nine nations with incomes above $15,000, or nations usually included among lists of "developed" countries. Slovenia, from the former Yugoslavia, is among the short list of six countries with incomes between $10,000 and $15,000, or countries sometimes listed as either developed or emerging.

Comparisons of per capita GNI is necessarily tied to world prices and to exchange rates among world currencies. The per capita GNI of Sierra Leone, for example, reflects how many products an average citizen could buy on the international market, and in this case very few, for its per capita GNI is only $190. On the basis of per capita GNI, around seventeen countries have incomes between $5,000 and $10,000. A few of those are growing rapidly, and may soon cross the $10,000 threshold, particularly the Czech Republic, Hungary, and Mexico. All the roughly 130 countries with incomes below $5,000, or less than an eighth of the income in the United States, are relatively poor, but among the nineteen with incomes over $3,000 are some major world powers, including Russia, Turkey, South Africa, and Brazil. The remaining 110 countries, all with incomes under $3,000, include over two-thirds of the world's total population, for China ($1,500) and India ($620) are among them, as well as such other populous countries as Indonesia, Pakistan, Bangladesh, Nigeria, and Sudan. At the very bottom are those fifty or so countries with incomes below $500.

Yet, the ranking of countries on the basis of GNI can be very misleading. This is obvious when one tries to determine how anyone could survive in the approximately thirty-five countries with incomes of less than $400 a year, or in the lowest of all. Burundi, on $90 a year. In the United States, a person could not survive for a week with that income. Thus, today, the fairest and increasingly most often cited income figure is what is called the Purchasing Power Parity income, or what I will refer to as PPP. Instead of currency exchange rates, this is based on a survey of the cost of hundreds of goods and services in the local currency. It includes statistical conversions that come as close as possible to estimating the real income among countries. In poor countries, generally, the cost of local food stuffs, and above all of human services, tends to be very low, unbelievably low when translated into dollars. In terms of local purchasing power and living standards, such currencies are drastically undervalued in exchange rates. The PPP corrects for this, and for most poor countries it is as much as five times higher than the per capita GDP. For example, in India the GDP in 2004 was only $620, but the PPP was $3,100; in booming China the GDP was $1,500, while its PPP was $5,890. At the higher incomes, the changes from GDP to PPP are small, and 111 some cases the PPP lower (dramatically so for Norway and Switzerland). But as one moves down the GDP, the gap between GDP and PPP becomes more pronounced.

In 2004, approximately thirty countries had a PPP over $20,000 (only twenty of these had a population of over 1 million, but notably, some tiny countries are among the most wealthy, with Luxemburg always at the top). These most affluent countries generally duplicate the present thirty countries that are members of the Organization of Economic Cooperation and Development (OCED), but not exactly, since some lower income countries (Mexico, Turkey) are in this elite organization. Twenty-two additional countries, some very small, had a PPP of over $10,000, or roughly the poverty level for a single person in the United States. This means that their living standards range from one-fourth to one-half that of the United States (which has a PPP of just under $40,000). These two groups total just over 1.222 billion, or barely 18.5 percent of the world's total population. At least most of these are generally listed as developed or industrialized countries. The exception would be a few countries that gain a high rank in incomes only because of the exportation of oil. But some countries almost always listed among industrialized or developed countries are not in the above $10,000 PPP group, and this includes Russia. Also note that much of this data is based

upon the self-reporting of countries, and may slightly overestimate incomes. Also, in some nations, particularly oil-rich countries, incomes are so skewed toward a few at the top as to leave the great mass of citizens at very low incomes.

Around forty countries have a PPP of $5,000 to $10,000. This means that living standards range from one-eighth to one-fourth of those in the United States. Some of these countries are often listed as having emerging economies, for some may soon cross the threshold of $10,000 (Russia, Mexico, Brazil, Thailand, and Turkey are the best candidates). Other quite populous countries in tins list (Philippines, Ukraine) are far from this goal of $10,000, as are the 1.3 billion people of China, who have PPP incomes near $6,000. These forty countries, ranging from near poor to emergent, contain 2.34 billion people, or 36 percent of the world's population (over half in China).

Almost 3 billion people have PPP incomes of under $5,000 in ninety-two countries, some very small. Whatever the euphemisms used to describe them (such as underdeveloped), those countries are simply poor. Most have little early prospects of moving above $5,000 PPP. But even here, those close to $5,000 are worlds apart from those at the lowest level. None of these aspirants are in sub-Saharan Africa, Albania, Armenia, El Salvador, and Paraguay are above $4,500, with Egypt, Guatemala, Jamaica, Jordan, Morocco, and Surinam above $4,000. India is a special case. Its 1.08 billion people nuke up over a third of those with incomes below $5,000. By its own accounting, its PPP had risen to $3,100 in 2004, and its annual growth rate is very high. It is conceivable that, in another decade or so, it will reach the $5,000 level. A total of forty countries have incomes between $2,000 and $5,000, while fifty-two countries are below $2,000. These make up the poorest of the poor, with thirty-two in sub-Saharan Africa, about a dozen in Asia, and only Haiti in the Western Hemisphere. The others are small island republics. Fifteen African countries are at $1,000 or below (Sierra Leone and Somalia are at the bottom at $600). The two most populous African countries—Nigeria and Ethiopia—are in this group. It is difficult to conceive of people surviving on one-fortieth the average purchasing power of Americans. Compounding the problem is that incomes in these poorest countries are often concentrated in small elites.

Average PPP levels may be very misleading if one is concerned about the overall welfare of a population. Vitally important is income distribution, or the degree of income equality. The best indicator of the general welfare of a population might be the average per capita PPP of those who suffer the lowest 20 percent of incomes. Unfortunately, income inequality is usually greater in poor countries than it is in affluent ones (see figure 3). Welfare includes several factors, some not tied to income. These include life expectancy, low infant mortality rates, universal access to health care and education, political stability, low crime rates, access to work, gender equality, and clean air and water. In the last decade, not only has the gap between affluent and poor countries widened, but even in most industrial countries income inequality has risen at frightening rates. Worst of all, among wealthy nations, is the United States, in which, in 2000, over 30 percent of all income went to the top 10 percent, only 1.8 percent to the bottom 10 percent. This has worsened since 2000. Such income inequality is one reason why, in most

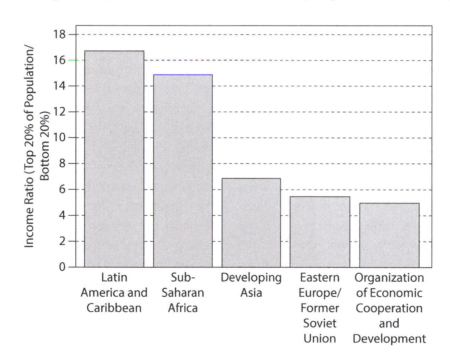

FIGURE 3. INCOME INEQUALITY WITHIN REGIONS.

Data from World Bank. United Nations, *Global Challenge, Global Opportunity*, 7.

attempts to measure overall welfare, the United States ranks below most western European countries.

The PPP can also miss some economic bases of a good life. Most services performed by homemakers and mothers do not make it into these accounts. Local barter transactions and black market sales (huge in some countries) are not counted. Thus, particularly in the poorest countries, even the PPP may somewhat underestimate the actual level of consumption.

In one critical way, national accounting indices almost always overstate the actual level of material welfare. This is what most concerns environmentalists. No such official estimates now include natural capital, and thus the environmental costs of production and consumption. Because of this, the present levels of income are not sustainable over the long term in most poor countries, and not even in wealthy countries without major substitutions of new types of energy and more drastic controls over pollution. Already, increasing amounts of capital and labor have to be devoted to environmental repair work, or the product of past, often reckless use of resources and the pollution of air and water. In some underdeveloped countries, if one deducted the annual loss of soil, of forest cover, and of endangered species, then the sustainable PPP might move toward zero. In highly developed and wealthy countries, such as the United States, the loss of natural capital might be only a small percentage of the total. But the lack of more specificity about such costs simply reflects the enormous difficulty of measuring the monetary value of natural capital. For example, if one assumes that global temperatures rise by 5°C in the next century, then what will be the cost to the world economy? It could be vast, but we will not soon know enough to even come close to a firm estimate. And what should the United States, which contributes almost one-fourth of the greenhouse gases that are helping

produce such warming, deduct from its present GDP in order to more realistically document its real annual income in sustainability terms? No one knows.

The United Nations maintains a System of National Accounts as part of its accounting division. In 1993, as a direct response to policies adopted at the Earth Summit in Rio in 1992, the United Nations tried to find a way to incorporate environmental costs into national accounting. This would have meant a revision in such categories as GNI or GDP. The problem was staggering in its complexity, and no revision resulted, but it did lead to a recommendation that the United Nations at least work toward such a goal and add supplemental or satellite accounts to address the environmental issue. Since then, economists have developed various strategies to gain a new Sustainable National Income index. In September 2005, the World Bank suggested that national accounts include certain natural resources, which often make up the largest share of wealth in poor countries. Unfortunately, it is these countries that arc most rapidly exporting, or using up, their natural assets.

Few would deny that the present national accounting is outdated and misleading, but so far the problem has remained a very complex, highly technical, academic enterprise, with various contenting factions. The most limited addition would involve only natural resources that already have market prices attached to them, such as minerals, oil, and even soil China, facing enormous environmental challenges, is the first country that is now planning a type of national accounting that includes environmental assets. The larger problem is accounting for life-supporting resources that are still part of the commons, such as oceans, streams, lakes, or the atmosphere. Also, how could a system of national accounts include global environmental costs. How could the United States calculate, and how should it pay for, its contribution to the depletion of the ozone layer, to ocean pollution, even to the acid rain that falls on Canadian forests and lakes? How could wealthy importing countries account for the environmental degradation they cause in the underdeveloped producing country? What is to keep wealthy countries from importing products that create environmental hazards abroad or exporting their own toxic wastes? Such issues implicate a system of international accounts that would, in proportionate ways, have to be incorporated into national accounts, and one can sense the level of international conflict that might ensue if the United Nations or regional agencies tried to implement such a system.

Equally difficult is any way of pricing the esthetic benefits of pristine forests and lakes, the opportunity to visit wilderness areas, the ability to enjoy threatened species of birds. One might poll citizens and find out how much they would be willing to pay in taxes each year to preserve such amenities, and on this basis give a monetary value to them. Or one might assume that present governmental resource regulations and reserves, and the costs that go with them, roughly indicate the value its citizens place on such amenities. But such measurements are meaningless in poor countries that cannot afford such taxes or cannot enforce environmental regulations. To the extent that the people of one country value habitats and biodiversity in other areas of the world, particularly poor areas, they should be willing to contribute to the foreign aid needed for their preservation.

## THE SPECIAL ENVIRONMENTAL CHALLENGES FACED BY POOR COUNTRIES

For the vast majority of humans, who live at a near subsistence level, such issues as sustainability must seem remote. Their great imperative has to be higher rates of productivity and increased consumption, whatever the long-term environmental costs. Given the large and usually still growing population in poor countries, this is an intimidating goal. Already, the assaults on local environments have too often reached crisis proportions, so much so that any sustained growth in incomes seems almost impossible. Yet, the assets available, in human capital, in sources of financing, in available tools of production, are meager at best, save for India and China, both of which have a large, university-trained workforce. Although most poor countries have tried to deal with environmental problems, and in some cases have had to do so, they have lacked the resources, or the political legitimacy, to enforce environmental legislation.

It is difficult to see how poor countries can alleviate most of these problems in the near, or even the distant, future. These run the gamut of environmental pressures: depleted fuels, eroded land, disappearing forest cover, threatened or already extinct species, uncontrolled urban growth, polluted air and water, and a scarcity of potable water or water for irrigation. Such poor countries have few tools to cope with such problems. Burdened by debts, by low market prices for exportable commodities, by political instability, they are all but helpless without major subsidies from wealthy countries. In part because of the highly subsidized and protected agriculture in the wealthy countries, they cannot even sell most food products on the world market. To make the situation even more cruel, they simply will not be able to retrace the paths of economic growth followed by Europe, America, and Japan. Not enough easily available or inexpensive resources remain for them to do so. And even if they try, environmental constraints will soon halt such growth—often not by any fault of their own, but because of the legacy left by rapid growth elsewhere.

Since 1850, and at an accelerating pace since 1950, the industrialized nations have attained a level of production of goods and services undreamed of in the human past. In 2000, a census year, the United States illustrated this consumptive largesse. It enjoyed a GNI of nearly $10 trillion (1996 dollars). In the previous year, its citizens had a disposable (after depreciation of capital and taxes) income of close to $6 trillion. Its agriculture was so productive that Americans had to spend just over 8 percent of this income for food used at home, or what costs over 50 percent of income in some poor countries. It spent over twice this amount for housing and household operation and on medical care, a third more on its automobiles, and an eighth more for recreation. Americans even spent over $80 billion on jewelry and personal care products, such as cosmetics. Each American spent, on average, about $2,300 on recreation, or almost as much as the PPP of India.

One key to this increased productivity has been the replacement of muscle power by other forms of energy, mainly from the controlled burning of fossil fuels. Another necessary condition has been the development of new tools and techniques of production, and thus ever greater efficiency. This has paid off most of all in agriculture, by far the most efficient sector in affluent countries today (the smallest input of labor for the output). Increasingly, the sources of fuels, timber, minerals, and tropical foods to sustain this rapid growth have been the less developed areas of the world. For example, Americans have not only drawn down their resources, such as oil and gas, but those of the world as a whole. Wealthy countries less favored by natural resources, such as Japan and Britain, have been almost completely dependent on such foreign resources. Poor countries have not been in a position to resist the out-shipment of precious resources, the loss of their natural capital. They have depended on the income for growing populations. Without a market for unprocessed goods, few underdeveloped

countries could begin to support their existing population. They are dependent either on trade or aid. One can only wonder what Nigeria will do when its oil runs out, given its engorged population and underdeveloped agriculture.

Almost all poor countries have tried to follow development patterns that succeeded in the wealthy countries, but only a few have done enough to improve agricultural production. Most have tried to introduce labor- intensive forms of manufacturing, with textiles often in the lead. And, indeed, as measured by present national income indices, most former colonial nations have enjoyed at least low levels of economic growth. Foreign aid and the green revolution have kept such economies growing. At least a minority of the population has benefitted, with a degree of affluence easily visible in favored areas of growing cities. Whether overall human welfare has improved is impossible to measure, but one can harbor doubts that it has.

What can poor nations do to move toward Western living standards? They cannot emulate the past history of the most industrialized nations. Few have the needed resources at home, and they cannot afford to import them, particularly energy. Population pressures have already decimated forests, eroded land, and exhausted local supplies of fuel even for cooking. Hungry peasants encroach upon parks and wildlife preserves, desperate for land, wood, or game. Others have overfished increasingly polluted streams, even as industrialized nations have joined in overfishing half the world's oceans. Agricultural reform might improve food production, but only by displacing most near-subsistence farmers. Few poor countries have the money or credit or needed skills to develop profitable manufacturing, and thus have to depend on foreign capital. Cheap labor is often their lure for foreign investment. Tropical countries have to export a few commercially important foodstuffs, or their dwindling reserves of timber, oil, gas, or minerals.

Increasingly, sub-Saharan Africa depends upon imported foods. Such countries cannot afford to attend to developing environmental problems.

In the last two decades, the dominant environmental concerns in the wealthy nations have shifted toward global issues. This largely involves the past role of industrialized societies in creating the problems that are now manifest in the underdeveloped world, and their responsibility in mitigating such global problems as atmospheric warming and, with it, climate change; tropical deforestation and, with it, a loss of biodiversity; and the likelihood of increased famines because of degraded soils and scarcities of water and energy. In a sense, the great overarching problem for poor countries is continued rapid population growth, a problem that wealthy countries, with stable or declining populations, can do little to influence from a distance.

## ENVIRONMENTAL MITIGATION IN WEALTHY COUNTRIES

In most respects, the citizens of the twenty most wealthy countries now enjoy the fruits of a century of environmental mitigation. As they look at their immediate surroundings, they see problems aplenty, but nothing to compare to the even recent past. Only forty years ago, people everywhere had to fear the effects of nuclear fallout from the testing of over two thousand bombs by cold war antagonists. No more, although some still fear accidents at nuclear reactors. In the early twentieth century, European and American cities, in the winter, suffered horrible smog and soot from the almost universal heating by coal. No more, although the continued use of coal for electrical generation contributes to smog, acid rain, and global warming. A century ago almost all human and industrial waste was dumped, untreated, in rivers and oceans. No more in wealthy countries, although problems of waste disposal still haunt them.

In the early twentieth century, farmers in the United States were still clearing forests and increasing the land cultivated, while erosion, by wind and water, was rife (think of the dust bowl or the red hillsides of the Southeast). No more. Despite a tripling of population in the United States since 1900, and a fourfold increase in agricultural production, the amount of cultivated land has declined by a third, and the forest cover is now more extensive than in 1900. Few now remember the ugliness of factory towns, or the tenements of large cities, or the unpainted shacks of share croppers, or when city streets ran with the manure and urine of horses. Epidemics of water-borne diseases (typhoid, cholera) were still present in 1900. Whales were becoming endangered species because of overhunting, while the last passenger pigeon died in a Cincinnati zoo in 1914. Americans came close to killing their last bison. Lead was a basic ingredient of paints, and soon also of gasoline. I could go on and on, if needed, to prove that for wealthy countries, from many environmental perspectives, the past was far worse than the present. But who doubts that fact?

The people of the underdeveloped world rightly envy the prosperity of the industrialized world, which is beyond their reach. They also value its ability to cope with environmental problems, which is even more beyond their reach. In so many areas, environmental regulations and new technologies of production have mitigated environmental problems in spite of increases in population and an even more rapid increase in per capita consumption. Not completely, of course, as present American realities testify.

Americans use energy more efficiently than in the past. They have reduced the carbon intensity of fuel as they moved from wood, to coal, to oil, and to gas (the generation of electricity by natural gas emits 50 times less carbon dioxide, per unit of heat, than a wood-burning stove). Yet, they still emit more carbon than ever before. This is a function of increases in population and, even more, consumption Americans have reduced the emission of sulfur compounds, and in a very limited way nitrous oxides, but they still suffer from air pollution and acid rain. They have reduced most of the risks of waste disposal, but they are overwhelmed with its volume. By great effort, they have saved many endangered species from extinction, but some habitat losses have effectively ended any but isolated enclaves. They have increased, at a rate much higher than population growth, the amount of land preserved in parks and wilderness areas, national and state forests, and wildlife sanctuaries, but nothing can relieve the pressure of more and more visitors to such sites, a function of numbers, affluence, cheap transportation, and a much more widespread public appreciation of the outdoors. At great effort and great cost, they have improved air quality in most areas of the country, and water quality in some areas, but again the sheer numbers and a growing use have imperiled precious aquifers and threatened overused streams. Because of a thousand new chemicals, and new products, they have pushed at least trace amounts of new compounds into the atmosphere, often with unknown risks. But as a whole, environmentally caused human mortality seems to be at the lowest level in human history in developed countries, no mean achievement (at the same time, mortality rates for many other species have gone up).

Behind the achievement was a transformation in human values. In the United States, every poll reveals broad public support for environmental legislation, although not for radical or sacrificial legislation. In much of western Europe, the public support is even higher. People take extra effort to recycle waste products, fight new urban developments in order to save forests, and are deeply concerned about threatened species. Of course, it is easier to the concerned when the problem is at a distance or when local costs are low. But the gap in understanding, in attitudes, between present Europeans and Americans and those of a century ago is deep. It is difficult to understand people who shot bison for the fun of it, who killed songbirds en masse, who gained a bounty for every wolf killed, who almost reflexively killed any snake observed, who saw trees as obstacles to progress, or who casually dumped wastes into the nearest stream.

But the high-consumption countries, in a global perspective, have incurred a high cost for their income and even their environmental repairs. They have used up a large share of the easily procured fossil fuels, threatened the protective ozone layer by their emission of ozone-depleting gases, risked a rapid rise in global temperatures because of greenhouse gas emissions, pushed the level of pollution in oceans to dangerous levels, inundated themselves with waste, used enormous quantities of water, exploited the most easily mined of the world s mineral resources, and, in earlier centuries, in both Europe and America, stripped away a large share of forest cover. By their excess, they have created an earth that can no longer support the type of development, in poor countries, that has led to their consumptive excesses. They got there first, and took the best.

# A NEW WORLD

By Bill McKibben

Imagine we live on a planet. Not our cozy, taken-for-granted earth, but a planet, a real one, with melting poles and dying forests and a heaving, corrosive sea, raked by winds, strafed by storms, scorched by heat. An inhospitable place.

It's hard. For the ten thousand years that constitute human civilization, we've existed in the sweetest of sweet spots. The temperature has barely budged; globally averaged, it's swung in the narrowest of ranges, between fifty-eight and sixty degrees Fahrenheit. That's warm enough that the ice sheets retreated from the centers of our continents so we could grow grain, but cold enough that mountain glaciers provided drinking and irrigation water to those plains and valleys year-round; it was the "correct" temperature for the marvelously diverse planet that seems right to us. And every aspect of our civilization reflects that particular world. We built our great cities next to seas that have remained tame and level, or at altitudes high enough that disease-bearing mosquitoes could not overwinter. We refined the farming that has swelled our numbers to take full advantage of that predictable heat and rainfall; our rice and corn and wheat can't imagine another earth either. Occasionally, in one place or another, there's an abrupt departure from the norm—a hurricane, a drought, a freeze. But our very language reflects their rarity: freak storms, disturbances.

In December 1968 we got the first real view of that stable, secure place. *Apollo 8* was orbiting the moon, the astronauts busy photographing possible landing zones for the missions that would follow. On the fourth orbit, Commander Frank Borman decided to roll the craft away from the moon and tilt its windows toward the horizon—he needed a navigational fix. What he got, instead, was a sudden view of the earth, rising. "Oh my God," he said. "Here's the earth coming up." Crew member Bill Anders grabbed a camera and took the photograph that became

Note: this chapter has been excerpted from the original publication.

the iconic image perhaps of all time. "Earthrise," as it was eventually known, that picture of a blue-and-white marble floating amid the vast backdrop of space, set against the barren edge of the lifeless moon. Borman said later that it was "the most beautiful, heart-catching sight of my life, one that sent a torrent of nostalgia, of sheer homesickness, surging through me. It was the only thing in space that had any color to it. Everything else was simply black or white. But not the earth." The third member of the crew, Jim Lovell, put it more simply: the earth, he said, suddenly appeared as "a grand oasis."

*But we no longer live on that planet.* In the four decades since, that earth has changed in profound ways, ways that have already taken us out of the sweet spot where humans so long thrived. We're every day less the oasis and more the desert. The world hasn't ended, but the world as we know it has—even if we don't quite know it yet. We imagine we still live back on that old planet, that the disturbances we see around us are the old random and freak-ish kind. But they're not. It's a different place. A different planet. It needs a new name. Eaarth. Or Monnde, or Tierrre, Errde, ОККУЧИВАТЬ. It still looks familiar enough—we're still the third rock out from the sun, still three-quarters water. Gravity still pertains; we're still earth*like*. But it's odd enough to constantly remind us how profoundly we've altered the only place we've ever known. I am aware, of course, that the earth changes constantly, and that occasionally it changes wildly, as when an asteroid strikes or an ice age relaxes its grip. This is one of those rare moments, the start of a change far larger and more thoroughgoing than anything we can read in the records of man, on a par with the biggest dangers we can read in the records of rock and ice.

Consider the veins of cloud that streak and mottle the earth in that glorious snapshot from space. So far humans, by burning fossil fuel, have raised the temperature of the planet nearly a degree Celsius (more than a degree and a half Fahrenheit). A NASA study in December 2008 found that warming on that scale was enough to trigger a 45 percent increase in thunderheads above the ocean, breeding the spectacular anvil-headed clouds that can rise five miles above the sea, generating "super-cells" with torrents of rain and hail. In fact, total global rainfall is now increasing 1.5 percent a decade. Larger storms over land now create more lightning; every degree Celsius brings about 6 percent more lightning, according to the climate scientist Amanda Staudt. In just one day in June 2008, lightning sparked 1,700 different fires across California, burning a million acres and setting a new state record. These blazes burned on the new earth, not the old one. "We are in the mega-free era," said Ken Frederick, a spokesman for the federal government. And that smoke and flame, of course, were visible from space—indeed anyone with an Internet connection could watch the video feed from the space shuttle *Endeavour* as it circled above the towering plumes in the Santa Barbara hills.

Or consider the white and frozen top of the planet. Arctic ice has been melting slowly for two decades as temperatures have climbed, but in the summer of 2007 that gradual thaw suddenly accelerated. By the time the long Arctic night finally descended in October, there was 22 percent less sea ice than had ever been observed before, and more than 40 percent less than the year that the Apollo capsule took its picture. The Arctic ice cap was 1.1 million square miles smaller than ever in recorded history, reduced by an area twelve times the size of Great Britain. The summers of 2008 and 2009 saw a virtual repeat of the epic melt; in 2008 both the Northwest and Northeast passages opened for the first time in human history. The first commercial ship to make the voyage through the newly opened straits, the MV *Camilla Desgagnes,* had an icebreaker on standby in case it ran into trouble, but the captain reported, "I didn't see one cube of ice."

This is not some mere passing change; this is the earth shifting. In December 2008, scientists from the National Sea Ice Data Center said the increased melting of Arctic ice was accumulating heat in the oceans, and that this so-called Arctic amplification now penetrated 1,500 kilometers inland. In August 2009, scientists reported that lightning strikes in the Arctic had increased twentyfold, igniting some of the first tundra fires ever observed. According to the center's Mark Serreze, the new data are "reinforcing the notion that the Arctic ice is in its death spiral." That is, within a decade or two, a summertime spacecraft pointing its camera at the North Pole would see nothing but open ocean. There'd be ice left on Greenland—but much less ice. Between 2003 and 2008, more than a trillion tons of the island's ice melted, an area ten times the size of Manhattan. "We now know that the climate doesn't have to warm any more for Greenland to continue losing ice," explained Jason Box, a geography professor at Ohio State University. "It has probably passed the point where it could maintain the mass of ice that we remember." And if the spacecraft pointed its camera at the South Pole? On the last day of 2008, the *Economist* reported that temperatures on the Antarctic Peninsula were rising faster than anywhere else on earth, and that the West Antarctic was losing ice 75 percent faster than just a decade before.

Don't let your eyes glaze over at this parade of statistics (and so many more to follow). These should come as body blows, as mortar barrages, as sickening thuds. The Holocene is staggered, the only world that humans have known is suddenly reeling. I am not describing what will happen if we don't take action, or warning of some future threat. This is the *current* inventory: more thunder, more lightning,

less ice. Name a major feature of the earth's surface and you'll find massive change.

…

So how did it happen that the threat to our fairly far-off descendants, which required that we heed an alarm and adopt precautionary principles and begin to take measured action lest we have a crisis for future generations, et cetera—how did that suddenly turn into the Arctic melting away, the tropics expanding, the ocean turning acid? How did time dilate, and "100 or 200 years from now" become yesterday?

The answer, more or less, is that global warming is a huge experiment. We've never watched it happen before, so we didn't know how it would proceed. Here's what we knew twenty years ago: the historic level of carbon dioxide in the atmosphere, the level that produced those ten thousand years of stability, was roughly 275 parts per million. And also this: since the dawn of the Industrial Revolution we'd been steadily increasing that total, currently raising it more than two parts per million annually. But no one really knew where the red line was—it was impossible to really know in advance at what point you'd cross a tripwire and set off a bomb. Like, say, melting all the ice in the Arctic.

The number that people tossed around for about a decade was 550 parts per million. Not because we had any real data showing it was the danger point, but because it was double the historic concentration, which made it relatively easy to model with the relatively crude computer programs scientists were using. One paper after another predicted what would happen to sea levels or forest composition or penguin reproduction if carbon dioxide levels doubled to 550 parts per million. And so— inevitably and insidiously—that's the number we fixated on. Since it wouldn't be reached until the middle of the twenty-first century, it seemed to offer a little margin; it meshed plausibly with political time, with the kind of gradual solutions leaders like to imagine. That is, a doubling of carbon dioxide would happen well beyond the time that anyone now in power was likely to still be in office, or still running the company. It was when everyone's *grandchildren* would be in charge. As late as 2004, the journalist Paul Roberts, in his superb book *The End of Oil,* was able to write quite correctly that "most climate models—indicate that once concentrations exceed 550 ppm we will start to witness 'dangerous' levels of warming and damage, especially in vulnerable areas, such as low-lying countries or those already suffering drought." But by then some doubt was beginning to creep in. Odd phenomena (large chunks of the Antarctic falling into the ocean, say) were unnerving scientists enough that, in Roberts's words, most "would much rather see concentrations stabilized at 450 ppm … where we might avoid most long-term effects and instead suffer a kind of 'warming light,' moderate loss of shorefront land, moderate loss of species, moderate desertification," and so on. And since even 450 was still 15 percent above our current levels, "we have a little room to breathe, which is handy."

Or would have been. But as it turns out, we had been like commentators trying to call an election on the basis of the first precinct to report. Right about 2005 the real returns began to flood in, *flood* being the correct verb. And what they showed was that those old benchmarks—550, 450—had been wishful thinking. No breathing room, not when hurricane seasons like 2005 were setting new records for insurance payouts, not when polar ice was melting "fifty years ahead of schedule," not when the tropics "appear to have already expanded during only the last few decades of the 20th century by at least the same margins as models predict for this century." Indeed, "ahead of schedule" became a kind of tic for headline writers: "Arctic Melt-off Ahead of Schedule" (the *Christian Science Monitor,* which quoted one scientist as saying "we're a hundred years ahead of schedule" in thawing Greenland), "Dry Future Well Ahead of Schedule" (the *Australian),* "Acidifed Seawater Showing Up Along Coast Ahead of Schedule" (the *Seattle Times).* The implication was that global warming hadn't read the invitation correctly and was showing up at four for the reception instead of six. In fact, of course, the "schedule" was wrong. And of course it was wrong—this was, as I've said, a huge experiment. Twenty-five years ago almost nobody even knew the planet was going to warm at all, never mind how fast.

It was that summer melt of Arctic ice in 2007 that seemed to break the spell, to start raising the stakes. The record minimums for ice were reached in the last week of September; in mid-December James Hansen, still the planet's leading climatologist, gave a short talk with six or seven slides at the American Geophysical Union meeting in San Francisco. What he said went unreported at the time, but it may turn out to be among the most crucial lectures in scientific history. He summarized both the real-world data that had emerged in recent years, including the ice-melt, and also the large body of research on paleoclimate—basically, the attempt to understand what had happened in the distant past when carbon dioxide levels climbed and fell. Taken together, he said, these two lines of inquiry made it clear that the safe number was, at most, 350 parts per million .

The day Jim Hansen announced that number was the day I knew we'd never again inhabit the planet I'd been born on, or anything close to it. Because we're already past 350—way past it. The planet has nearly 390 parts per million carbon dioxide in the atmosphere. We're too high. Forget the grandkids; it turns out this was a problem for our *parents.*

We can, if we're very lucky and very committed, eventually get the number back down below 350. This task will be extremely hard. The planet can, slowly, soak up excess carbon dioxide if we stop pouring more in. That fight is what I spend my life on now, because it's still possible we can avert the very worst catastrophes. But even so, great damage will have been done along the way, on land and in the sea. In September 2009 the lead article in the journal *Nature* said that above 350 we "threaten the ecological life-support systems that have developed in the late Quaternary environment, and severely challenge the viability of contemporary human societies." A month later, the journal *Science* offered new evidence of what the earth was like 20 million years ago, the last time we had carbon levels this high: sea levels rose one hundred feet or more, and temperatures rose as much as ten degrees. The Zoological Society of London reported in July 2009 that "360 is now known to be the level at which coral reefs cease to be viable in the long run."

We're not, in other words, going to get back the planet we used to have, the one on which our civilization developed. We're like the guy who ate steak for dinner every night and let his cholesterol top 300 and had the heart attack. Now he dines on Lipitor and walks on the treadmill, but half his heart is dead tissue. We're like the guy who smoked for forty years and then he had a stroke. He doesn't smoke anymore, but the left side of his body doesn't work either.

Consider: On January 26, 2009, less than a week after taking office, Barack Obama announced a series of stunning steps designed to dramatically raise fuel efficiency for cars. He also named a new envoy to aggressively negotiate an international accord on global warming. "This should prompt cheers from California to Maine," the head of one environmental group exulted. "The days of Washington dragging its heels are over," insisted the president. It was the most auspicious day of environmental news in the twenty years of the global warming era. And then that afternoon, the National Oceanic and Atmospheric Administration released a new study showing that a new understanding of ocean physics proved that "changes in surface temperature, rainfall, and sea level are largely irreversible for more than a thousand years after carbon dioxide emissions are completely stopped." Its author, Susan Solomon, was interviewed on National Public Radio that night. "People have imagined that if we stopped emitting carbon dioxide that the climate would be back to normal in one hundred years or two hundred years," she said. "What we're showing here is that that's not right." No one is going to refreeze the Arctic for us, or restore the pH of the oceans, and given the momentum of global warming we're likely to cross many more thresholds even if we all convert to solar power and bicycles this afternoon.

Which, it must be said, we're not doing. The scientists didn't merely underestimate how fast the Arctic would melt; they overestimated how fast our hearts would melt. The Intergovernmental Panel on Climate Change, or IPCC, carefully calculated a variety of different "emissions pathways" for the future, ranging from a world where we did everything possible to make ourselves lean and efficient to a "business-as-usual" model where we did next to nothing. In the last decade, "as the United States has done very little to change its energy habits, and as the large Asian economies have come online, carbon emissions have risen "far above even the bleak scenarios" considered in the reports. In the summer of 2008, at an academic conference at Britain's Exeter University, a scientist named Kevin Anderson took the podium for a major address. He showed slide after slide, graph after graph, "representing the fumes that belch from chimneys, exhausts and jet engines, that should have bent in a rapid curve towards the ground, were heading for the ceiling instead." His conclusion: it was "improbable" that we'd be able to stop short of 650 parts per million, even if rich countries adopted "draconian emissions reductions within a decade." That number, should it come to pass, would mean that global average temperatures would increase something like seven degrees Fahrenheit, compared to the degree and a half they've gone up already.

...

To give you an idea of how aggressively the world's governments are willing to move, in July 2009 the thirteen largest emitters met in Washington to agree on an "aspirational" goal of 50 percent cuts in carbon by 2050, which falls pretty close to the category of "don't bother."

...

So far we've been the cause for the sudden surge in greenhouse gases and hence global temperatures, but that's starting to change, as the heat we've caused has started to trigger a series of ominous feedback effects. Some are fairly easy to see: melt Arctic sea ice, and you replace a shiny white mirror that reflects most of the incoming rays of the sun back out to space with a dull blue ocean that absorbs most of those rays. Others are less obvious, and much larger: booby traps, hidden around the world, waiting for the atmosphere to heat.

For instance, there are immense quantities of methane—natural gas—locked up beneath the frozen tundra, and in icy "clathrates" beneath the sea. Methane, like carbon dioxide, is a heat-trapping gas; if it starts escaping into the atmosphere, it will add to the pace of warming. And that's what seems to be happening, well ahead (need it be said) of schedule. In 2007, atmospheric levels of methane began to spike. Scientists weren't sure where they were coming from, but the fear was that those tundra and ocean sources were starting to melt in earnest. In the summer of 2008, a Russian research ship,

the *Jacob Smirnitskyi*, was cruising off the country's northern coast in the Laptev Sea when the scientists on board started finding areas of the water's surface foaming with methane gas. Concentration's were a hundred times normal: "Yesterday, for the first time, we documented a field where the release was so intense that the methane did not have to dissolve into the sea water but was rising as methane bubbles to the sea surface," one of the scientists e-mailed a journalist at the *Independent*. "These methane chimneys were documented on an echo sounder and with seismic instruments." The head of the research team, Igor Semiletov of the University of Alaska in Fairbanks, noted that temperatures over eastern Siberia had increased by almost ten degrees in the last decade. That's melting permafrost on the land, and hence more relatively warm water is flowing down the region's rivers into the ocean, where it may in turn be melting the icy seal over the underwater methane. The melting permafrost is also releasing methane on land. "On helicopter flights over the delta of the Lena River, higher methane concentrations have been measured at altitudes as high as 1,800 meters," reported Natalia Shakhova, of the Russian Academy of Sciences. In recent winters scientists have reported that far northern ponds and marshes stayed unfrozen even in the depths of winter because so much methane was bubbling out from underneath. "It looks like a soda can is open underneath the water," one researcher explained.

That's scary. Scarier even than the carbon pouring out of our tailpipes, because we're not directly releasing that methane. We burned the coal and gas and oil, and released the first dose of carbon, and that raised the temperature enough to start the process in motion. We're responsible for it, but we can't shut it off. It's taken on a life of its own. One recent estimate: the permafrost traps 1,600 billion tons of carbon.

A hundred billion tons could be released this century, mostly in the form of methane, which would have a warming effect equivalent to 270 years of carbon dioxide emissions at current levels. "It's a kind of slow-motion time bomb," said Ted Schuur of the University of Florida in March 2009. At a certain point, he added, "the feedback process would continue even if we cut our green house emissions zero"

We don't know if methane release has begun in earnest yet, or the exact threshold we'd need to pass. But there are dozens of such feedback loops out there. Peat covers about 2 percent of the planet's land surface, mostly in the far north—think moors, bogs, mires, swamp forests. They are wet places filled with decaying vegetation, a kind of nursery for what in many millennia could become coal. Because they're wet, they're very stable; the plants decompose very very slowly, so peatlands make a perfect: "sink" for carbon, holding perhaps half as much as the atmosphere. But say you raise the temperature and hence the rate of evaporation; the water table starts to fall, and those swamps start to dry out. And as they do, the carbon in all that decaying vegetation starts to decompose more quickly and flood into the atmosphere. A 2008 study found, in fact, that "peatlands will quickly respond to the expected warming in this century by losing labile soil organic carbon during dry periods." How much? Well, peat bogs worldwide hold the equivalent of sixty-five years of fossil-fuel burning, and the expected warming will dry out enough of them to cause the loss of between 40 and 86 percent of that carbon. It's as if we'd conjured up out of nowhere a second human population that's capable of burning coal and oil and gas nearly as fast as we do.

At the same time that we're triggering new pulses of carbon into the atmosphere, we're also steadily weakening the natural systems that pull it out of the air. Normally—over all but the last two hundred years of human civilization—the carbon dioxide level in the atmosphere remained stable because trees and plants and plankton sucked it up about as fast as volcanoes produced it. But now we've turned our cars and factories into junior volcanoes, and so we're not just producing carbon faster than the plant world can absorb it; we're also making it so hot that the plants absorb less carbon than they used to. In a 2008 experiment, scientists carved out small plots of grassland and installed them in labs where they could heat them artificially. "During this anomalously warm year and the year that followed, the two plots sucked up two-thirds less carbon-than the plots that had been exposed to normal temperatures," the researchers reported. The same thing may be happening at sea, where in January 2009 scientists "issued a warning" after finding "a sudden and dramatic collapse in the amount of carbon emissions absorbed" in fast-warming areas of the Sea of Japan. Imagine that you desperately need to bail out your boat, but you find that your buckets are filled with holes that keep getting larger. "Fifty years ago, for every ton of $CO_2$ emitted to the atmosphere, natural sinks removed 600 kilograms. Currently the sinks are removing only 559 kilograms per ton, and the amount is falling." Those are big holes.

. . .

# Section II

## SYSTEMS THINKING AND THE LIMITS TO GROWTH

# THE SYSTEM LENS

By Donella H. Meadows

Managers are not confronted with problems that are independent of each other, but with dynamic situations that consist of complex systems of changing problems that interact with each other. I call such situations messes. … Managers do not solve problems, they manage messes.

—Russell Ackoff, operations theorist

Early on in teaching about systems, I often bring out a Slinky. In case you grew up without one, a Slinky is a toy—a long, loose spring that can be made to bounce up and down, or pour back and forth from hand to hand, or walk itself downstairs.

I perch the Slinky on one upturned palm. With the fingers of the other hand, I grasp it from the top, partway down its coils. Then I pull the bottom hand away. The lower end of the Slinky drops, bounces back up again, yo-yos up and down, suspended from my fingers above.

"What made the Slinky bounce up and down like that?" I ask students.

"Your hand. You took away your hand," they say.

So I pick up the box the Slinky came in and hold it the same way, poised on a flattened palm, held from above by the fingers of the other hand. With as much dramatic flourish as I can muster, I pull the lower hand away.

Nothing happens. The box just hangs there, of course.

"Now once again. What made the Slinky bounce up and down?"

The answer clearly lies within the Slinky itself. The hands that manipulate it suppress or release some behavior that is latent within the structure i of the spring.

That is a central insight of systems theory.

Once we see the relationship between structure and behavior, we can begin to understand how systems work, what makes them produce poor results, and how to shift them into better behavior patterns. As our world continues to change rapidly and become more complex, systems thinking will help us manage, adapt, and see the wide range of choices we have before us. It is a way of thinking that gives us the freedom to identify root causes of problems and see new opportunities.

So, what is a system? A system is a set of things—people, cells, molecules, or whatever—interconnected in such a way that they produce their own pattern of behavior over time. The system may be buffeted, constricted, triggered, or driven by outside forces. But the system's response to these forces is characteristic of itself, and that response is seldom simple in the real world.

When it comes to Slinkies, this idea is easy enough to understand. When it comes to individuals, companies, cities, or economies, it can be heretical. The system, to a large extent, causes its own behavior! An outside event may unleash that behavior, but the same outside event applied to a different system is likely to produce a different result.

Think for a moment about the implications of that idea:

- Political leaders don't cause recessions or economic booms. Ups and downs are inherent in the structure of the market economy.
- Competitors rarely cause a company to lose market share. They may be there to scoop up the advantage, but the losing company creates its losses at least in part through its own business policies.
- The oil-exporting nations are not solely responsible for oil-price rises. Their actions alone could not trigger global price rises and economic chaos if the oil consumption, pricing, and investment policies of

the oil-importing nations had not built economies that are vulnerable to supply interruptions.
- The flu virus does not attack you; you set up the conditions for it to flourish within you.
- Drug addiction is not the failing of an individual and no one person, no matter how tough, no matter how loving, can cure a drug addict—not even the addict. It is only through understanding addiction as part of a larger set of influences and societal issues that one can begin to address it.

Something about statements like these is deeply unsettling. Something else is purest common sense. I submit that those two some things—a resistance to and a recognition of systems principles—come horn two kinds of human experience, both of which are familiar to everyone.

On the one hand, we have been taught to analyze, to use our rational ability, to trace direct paths from cause to effect, to look at things in small and understandable pieces, to solve problems by acting on or controlling the world around us. That training, the source of much personal and societal power, leads us to see presidents and competitors, OPEC and the flu and drugs as the causes of our problems.

On the other hand, long before we were educated in rational analysis, we all dealt with complex systems. We are complex systems— our own bodies are magnificent examples of integrated, interconnected, self-maintaining complexity. Every person we encounter, every organization, every animal, garden, tree, and forest is a complex system. We have built up intuitively, without analysis, often without words, a practical understanding of how these systems work, and how to work with them.

Modern systems theory, bound up with computers and equations, hides the fact that it traffics in truths known at some level by everyone. It is often possible, therefore, to make a direct translation from systems jargon to traditional wisdom.

Because of feedback delays within complex systems, by the time a problem becomes apparent it may be unnecessarily difficult to solve.
> —*A stitch in time saves nine.*

According to the competitive exclusion principle, if a reinforcing feedback loop rewards the winner of a competition with the means to win further competitions, the result will be the elimination of all but a few competitors.
> —*For he that hath, to him shall be given; and he that hath not, from him shall be taken even that which he hath (Mark 4:25)*
> *or*
> —*The rich get richer and the poor get poorer.*

A diverse system with multiple pathways and redundancies is more stable and less vulnerable to external shock than a uniform system with little diversity.
> —*Don't put all your eggs in one basket.*

Ever since the Industrial Revolution, Western society has benefited from science, logic, and reductionism over intuition and holism. Psychologically and politically we would much rather assume that the cause of a problem is "out there," rather than "in here." It's almost irresistible to blame something or someone else, to shift responsibility away from ourselves, and to look for the control knob, the product, the pill, the technical fix that will make a problem go away.

Serious problems have been solved by focusing on external agents—preventing smallpox, increasing food production, moving large weights and many people rapidly over long distances. Because they are embedded in larger systems, however, some of our "solutions" have created further problems. And some problems, those most rooted in the internal structure of complex systems, the real messes, have refused to go away.

Hunger, poverty, environmental degradation, economic instability, unemployment, chronic disease, drug addiction, and war, for example, persist in spite of the analytical ability and technical brilliance that have been directed toward eradicating them. No one deliberately creates those problems, no one wants them to persist, but they persist nonetheless. That is because they are intrinsically systems problems—undesirable behaviors characteristic of the system structures that produce them. They will yield only as we reclaim our intuition, stop casting blame, see the system as the source of its own problems, and find the courage and wisdom to *restructure* it.

Obvious. Yet subversive. An old way of seeing. Yet somehow new. Comforting, in that the solutions are in our hands. Disturbing, because we must *do things,* or at least *see things* and *think about things,* in a different way.

When our small research group moved from MIT to Dartmouth College years ago, one of the Dartmouth engineering professors watched us in seminars for a while, and then dropped by our offices. "You people are different," he said. "You ask different kinds of questions. You see things I don't see. Somehow you come at the world in a different way. . . . How? Why?"

## INTERLUDE • *The Blind Men and the Matter of the Elephant*

Beyond Ghor, there was a city. All its inhabitants were blind. A king with his entourage arrived nearby; he brought his army and camped in the desert. He had a mighty elephant, which he used to increase the people's awe.

The populace became anxious to see the elephant, and some sightless from among this blind community ran like fools to find it.

As they did not even know the form or shape of the elephant, they groped sightlessly, gathering information by touching some part of it.

Each thought that he knew something, because he could feel a part. . . .

The man whose hand had reached an ear . . . said: "It is a large, rough thing, wide and broad, like a rug."

And the one who had felt the trunk said: "I have the real facts about it. It is like a straight and hollow pipe, awful and destructive."

The one who had felt its feet and legs said: "It is mighty and firm, like a pillar."

Each had felt one part out of many. Each had perceived it wrongly. . . .[2]

This ancient Sufi story was told to teach a simple lesson but one that we often ignore: The behavior of a system cannot be known just by knowing the elements of which the system is made.

. . . I don't think the systems way of seeing is better than the reductionist way of thinking. I think it's complementary, and therefore revealing. You can see some things through the lens of the human eye, other things through the lens of a microscope, others through the lens of a telescope, and still others through the lens of systems theory. Everything seen through each kind of lens is actually there. Each way of seeing allows our knowledge of the wondrous world in which we live to become a little more complete.

At a time when the world is more messy, more crowded, more interconnected, more interdependent, and more rapidly changing than ever before, the more ways of seeing, the better. The systems-thinking lens allows us to reclaim our intuition about whole systems and

- hone our abilities to understand parts,
- see interconnections,
- ask "what-if" questions about possible future behaviors, and
- be creative and courageous about system redesign.

Then we can use our insights to make a difference in ourselves and our world.

# THE HISTORY OF THE LIMITS TO GROWTH

By Jørgen Stig Nørgård, John Peet, and Kristín Vala Ragnarsdóttir

A pioneering report, *The Limits to Growth*, published in 1972, marked a turning point in thinking about the environment, selling some 30 million copies in 30 languages.[1] The two-year study behind the report took place at the Massachusetts Institute of Technology at the request of the Club of Rome, an international group of distinguished business people, state officials, and scientists founded by Aurelio Peccei, a former Fiat executive and president of Olivetti. Their concerns about the consequences of unrestrained growth in global population, resource consumption, and pollution led them to contact Jay W. Forrester, a professor in management at MIT, who had developed a method for analyzing the behavior of complex systems by means of simple simulation models. Forrester accepted the challenge and assembled a team of young experts, headed by Dennis Meadows. Meadows and his team constructed a model, known as World3, to keep track of the development of the study's central

parameters and their interactions. *The Limits to Growth (LtG)* was based on analyses found in the project's more detailed report.[2] Coauthored by Dennis Meadows, his biophysicist wife, Donella Meadows, the Norwegian management expert Jørgen Randers, and William Behrens III, it was designed to disseminate the group's findings to the broader public. The authors remained involved with the issues raised by *LtG* in the following decades, and Donella Meadows in particular was highly engaged, as well as being the most optimistic of the original authors, until her death in 2001.

The report's argument that the biosphere has a limited ability to absorb human population growth, production, pollution, and economic growth in general stirred considerable debate. Over the ensuing decades, however, a cohort of critics managed to derail the debate, apparently because they simply could not imagine that two centuries of impressive growth in Western economic production and consumption could ever run into any limits.

The renewed interest in problems such as global warming and economic crisis suggests that it is time to revive the derailed discussion about economic growth and the environment and to reconsider future development. In re-examining the analysis and central arguments of LtG, we have found that its approach remains useful and that its conclusions are still surprisingly valid.

Most environmentalists are familiar with the arguments made in LtG, but unfortunately the report has been largely dismissed by critics as a doomsday prophecy that has not held up to scrutiny.

Matthew R. Simmons, president of the world's largest investment company specializing in energy, Simmons and Company International, read the book a few years ago, after hearing about the controversy. To his surprise he discovered that the criticisms had little to do with the content of the book. "After reading Limits to Growth, I was amazed," he wrote in 2000. "There was not one sentence or even a single word written about an oil shortage or limits to any specific resource, by the year 2000."[3] He concluded that LtG broadly gives a correct picture of world development, and he became upset that so many of his colleagues had wasted three decades criticizing it instead of taking action.

## VARIOUS SCENARIOS OF LtG ANALYSES

What prompted the attack on LtG? In the book's conclusion, the authors assert: "If the present growth trends in world population, industrialization, pollution, food production, and resource depletion continue unchanged, the limits to growth on this planet will be reached sometime within the next one hundred years. The most probable result will be a rather sudden and uncontrollable decline in both population and industrial capacity."[1]

Aided and abetted by a drama-hungry media, this *one* scenario was seized upon by critics who sought to dismiss the book as another in a long line of hysterical doomsday prophecies. For example, many critics argued that the book did not give enough credence to human ingenuity and adaptability, which could prevent the collapse forecast in its model.[4] In addition, many economists claimed that the market mechanism, by adjusting prices accordingly, would lead to substitutions for scarce resources and would prompt inventors and entrepreneurs to develop various technological solutions, thereby preventing a collapse.[5] These critics focused on LtG's most dire warnings, ignoring some of the authors' more optimistic and constructive scenarios, such as the following: "It is possible to alter these growth trends and to establish a condition of ecological and economic stability that is sustainable far into the future. The state of global equilibrium could be designed so that the basic material needs of each person on earth are satisfied and each person has an equal opportunity to realize his individual human potential."[1] The sooner mankind changes its development path, the book concludes, the better are the chances of success.

Usually, research on the future consists of attempts to *predict* what will actually happen. Hence, the scenario methodology used in LtG of presenting *various* future options from which societies could choose appeared incomprehensible to many readers, who therefore paid attention only to the disastrous growth scenario. The message of the book is to point out the importance of changing course before causing irreversible damage to the environment and its life-support systems for billions of people. One of the authors of LtG, the futurist Jørgen Randers, has recently discussed the probability of such a collapse due to "the unfortunate combination of global decision delays and self-reinforcing feedback in the climate system."[6]

*LtG* should be given credit for emphasizing early on the interconnections and feedback between various sectors and trends. Today we see, for example, how our fast depletion of fossil fuel resources is directly contributing to climate change problems.

## Derailing the Debate

How was it possible to derail the *LtG* debate to the extent that the book and its message were essentially ignored (or, arguably, covered up) for decades?

One reason is that a book that hints at the necessity of curbing economic growth is very unwelcome to those who have a large stake in the status quo. This applies at the financial level, where *LtG* challenges many commercial interests in growth; at the political level, where governments fear dwindling tax revenue for public spending; and among professional mainstream economists, who instinctively resist a change in the paradigm of eternal growth and who have rarely addressed the question of how to plan for a steady-state economy in an orderly way.

A common pattern among *LtG*'s critics has been to erroneously charge that it predicted oil would run out before the year 2000, and then to write off the book by pointing out that this scenario did not actually happen.[7]

Among the most vigorous critics of *LtG* have been radical free-market economist Julian Simon and futurist Herman Kahn, who charged that the *LtG* models had "been damned as foolishness or fraud by every serious economic critic."[8] As proof, they cited Nobel laureate in economics Gunnar Myrdal. Myrdal's critique, however, turns out to be quite modest and not directed at the results of *LtG* but rather at its methodology. For instance, Myrdal wrote that

the authors'"use of mathematical equations and a huge computer" contributed little, if anything, to their scientific validity; the same conclusions could have been reached by "hard simple thinking" and an awareness "of the limitations of what we know."[8,9] The authors of *LtG* might have agreed with Myrdal,[10] but it is worth noting that prior to *LtG*'s appearance very few people had engaged in this sort of "hard simple thinking." In fact, Myrdal himself was skeptical of the growth-at-all-costs, GDP-obsessed approach of radical free-market economists such as Simon, arguing, for example, that this perspective neglected the value of supposedly nonproductive activities like leisure time.[9]

Continued growth is for many not an issue for debate; it is an indispensable condition of economic life. But there are many holes in their argument. Firstly, we have only just entered the century in which *LtG*'s growth scenario drama is supposed to unfold. None of the book's scenarios suggested that we would encounter significant difficulty before the period 2010–2030.

Secondly, out of all of the possibilities described in *LtG*, the actual global development trends observed so far have been in generally good agreement with the growth scenario, which assumed no major changes in the physical, economic, or social relationships (until the model later indicates collapse).[11,12]

Thirdly, we are already observing some imminent crises on the horizon, which in their global character remind us uncomfortably of *LtG*'s growth scenario. At least one type of pollution, the atmospheric concentration of $CO_2$, is now recognized as a serious threat to the earth's climate.[13] Similarly, experts foresee the supply of oil peaking within a decade.[14,15,16] The *LtG* computer model does not explicitly deal with $CO_2$ pollution or with oil consumption, but the development trends for these two parameters closely resemble trends for *total* pollution and *total* resource depletion in *LtG*'s growth scenario. So do many other environmental problems that we face today, including the increased levels of toxic substances in the oceans and in groundwater, the depletion of ocean fish stocks, and deforestation.

A different reason why *LtG* has been shelved for so long could be that the temporary peak in international oil prices that occurred after the OPEC-induced crises of the 1970s was soon followed by lower prices, suggesting that the outcomes outlined by *LtG* were easily corrected after a few years, courtesy of the marketplace. It was therefore tempting for those so inclined to conclude that apparently there were no physical limits to anything of relevance to economic growth. Hence the report was judged to be untrustworthy, although its modelling approach did not even consider this sort of short-term (primarily political) oscillation.

The fact that a relatively small part of the world has for a couple of centuries experienced exponential growth in material production and consumption is often used to argue that this trend can continue globally forever. This view conveniently ignores that most of humanity's tenure on earth has been accompanied by essentially zero growth.

## DEVELOPMENT WITHOUT GROWTH

Regrettably, the media did not devote much attention to *LtG*'s optimistic and interesting sketches of an achievable and sustainable global *development*, as opposed to continuous and unsustainable economic growth.[9] It is possible to have development both with and without growth, as illustrated by how a person's physical growth stops during the late teenage years while nonphysical development continues. Similarly, a society can continue to develop after growth stops. Once a society has "grown up," so to speak, "those pursuits that many people would

list as the most desirable and satisfying activities of man—education, art, music, religion, basic scientific research, athletics, and social interactions—could flourish."[1] A community without growth can be just as dynamic as our current growth economy, in that some branches decline and collapse while others sprout and grow within the limited framework that is always present in any overall economy—just like a centuries-old rainforest can have plenty of individual growth and decay while the whole remains stable.

What have Western societies been doing in the thirty-five years since the appearance of *LtG* and similar warnings from the same period?[17,18] When it comes to taking serious action, little has been done to reach a sustainable form of development, with the exception of some modest technical adjustments. On the other factors, such as population, consumption, and production, the political steps taken have generally been in the opposite direction from sustainability, and they have more than offset any benefits of technical progress. The result is that global environmental pressure, as for example indicated by humanity's ecological footprint,[19] is today much worse. And there are few signs of significant action toward changing this trend.

## BETTER GLOBAL DISTRIBUTION

Even though World3, the system-dynamics computer model behind the *LtG* assessment, handles the world as a whole, poor as well as rich, the authors explicitly stressed that the distribution of wealth and consumption plays a crucial role in real development. Until now, growth has not, as often promised, been used to reduce inequalities but rather to sustain a substantial gap between rich and poor, without having to deal with too much social unrest. By arguing that the economic cake cannot grow infinitely, *LtG* added moral legitimacy to those demanding more equality, both within nations and globally.

Globally, demand for more equality will imply that the slowdown or reduction in material consumption, which *LtG* recommends, has to start in the affluent countries. Halving the *global* environmental pressure would, for instance, require that affluent countries reduce their level of resource exploitation to only about one-tenth of the present level.[20] Arguments that Western countries need to have growth for the sake of the poor countries are revealed as empty rhetoric. It is the other way around. The affluent countries must hold back in order to ensure environmental space for those that need growth.[21]

One of the largest political fears concerning a steady-state society is unemployment combined with stagnating consumption. The obvious solution to ensuring jobs for all those capable of working would be to reduce work time by, for example, 20 percent instead of firing one-fifth of the work force. In 1935 the philosopher Bertrand Russell described the practice of firing employees to solve the problem of excess workers in the following manner: "In this way it is ensured that the unavoidable leisure shall cause misery all around instead of being a universal source of happiness. Can anything more insane be imagined?"[22] In spite of this, increasing unemployment by firing some people remains the established solution, together with efforts to increase consumption and production, which add to global environmental pressure.

## THE LIMITS TO GROWTH AS PART OF A SOLUTION

It would be unjust to describe all economists as unconditional supporters of eternal economic expansion. Throughout history economists have commonly considered growth as a temporary phase in the development of society.[23,24] But in the wake of World War II, growth—as defined by the new concept of Gross Domestic Product—became the unquestioned goal of politicians. Apart from a few ecologically oriented economic thinkers,[25,26] the majority of postwar economists viewed themselves as the prime and indispensable technicians of GDP maximization.

The recent renewed interest in the environment and economic development gives hope for a solution. Although it has not yet led to new action, this shift in thinking has triggered a few analyses that recognize possible limits to growth and hence point toward solutions along the lines suggested in *LtG*. The following examples illustrate this hope.

A recent UK government committee indicates an emerging political willingness to at least challenge the growth paradigm as reflected in the title of the committee's report: *Prosperity without Growth?* The report "questions whether ever-rising incomes for the already-rich are an appropriate goal for policy in a world constrained by ecological limits."[27,28]

Joseph Stiglitz, a Nobel Prize winner in economics who had at first rejected *LtG*'s ideas about resource shortages, now recognizes that present trends in the world economy are unsustainable.[29] Stiglitz, along with another Nobel laureate in economics, Amartya Sen, headed a commission convened by French president Nicolas Sarkozy to investigate alternative measures of social progress to GDP. One of their key messages is that "the time is ripe for our measurement system to shift emphasis from measuring economic production to measuring people's well-being."[30] In their critique of societies' overreliance on GDP, Stiglitz and Sen are implicitly agreeing with *LtG*'s analysis.

Finally, as a sign of renewed recognition of the limits to growth, 28 scientists have identified nine *planetary boundaries* within which human activities can operate safely. The scientists estimate that humanity has already transgressed three of these boundaries, namely those for climate change, biodiversity loss, and changes to the global nitrogen cycle.[31,32]

A sustainable future is not a matter of technology alone, partly because of the rebound effect.[33] It must build on new ways to live and organize societies, for instance recognizing that today "corporations have even less incentive than individuals to keep the Commons in order; in fact they have a (legal) clear line of responsibility to their shareholders alone and have continuously resisted government and international efforts to regulate the Commons."[34] This is referring to G. Hardin's classical paper on the problems of sharing the limited capacity of the Commons.[35] Elinor Ostrom won the 2009 Nobel Prize in Economics for her analysis of economic governance, and especially of the commons, that understands this conundrum. She shared the prize with Oliver A. Williamson, who was lauded for his analysis of economic governance, especially in regard to the boundaries of the modern corporation.[36] Both study how individuals can work together and share scarce resources, an ethos shared by the authors of LtG.[37]

The notion of a globalized world and calls for the continued deregulation of markets must be replaced by the firm democratic regulation of each economy, based on the best available scientific, technological, social, and moral and ethical information. The need for this has been reinforced by the recent financial crisis. In many researchers' opinions, a change of this nature will be better for both environmental and human well-being.

Governments wanting to undertake serious preparations for a transition toward a sustainable steady-state economy can, fortunately, draw on the many economists already experienced in the field. And it would be a pity in the process to discard the wisdom in the 1972 LtG and its later versions.[10,38] Clearly, the book has withstood the test of time and indeed, has only become more relevant.

## References

1. Meadows, DH, Meadows, DL, Randers, J & Behrens III, WW. The Limits to Growth (Universe Books, New York, 1972).

2. Meadows, DL et al. Dynamics of Growth in a Finite World (Wright-Allen Press Inc., Cambridge, USA, 1974).

3. Simmons, MR. Revisiting The Limits to Growth: Could the Club of Rome have been correct, after all? Part one. Energy Bulletin [online] (2000) (http://www.energybulletin.net/1512.html).

4. Jahoda, M. in Thinking about the Future: A Critique of "The Limits to Growth" (eds Cole, HSD, Freeman, C, Jahoda, M & Pavitt, KLR), Postscript on social change, 209–216 (Chatto & Windus for Sussex University Press, Sussex, UK, 1973).

5. Simon, JL. The Ultimate Resource (Princeton University Press, Princeton, 1981).

6. Randers, J. Global collapse—Fact or fiction? Futures 40 853–864 (2008).

7. Lomborg, B. The Sceptical Environmentalist: Measuring the Real State of the World, 121 (Cambridge University Press, UK, 2001).

8. Simon, JL & Kahn, H. The Resourceful Earth: A Response to Global 2000, 38 (Basil Blackwell, New York, 1984).

9. Myrdal, G. Against the Stream: Critical Essays on Economics, 204–205 (Pantheon Books, New York, 1973).

10. Meadows, DH, Randers, J & Meadows, DL. Limits to Growth: The 30-Year Update, xviii (Chelsea Green, White River Junction, VT, USA, 2004).

11. Turner, GM. A comparison of The Limits to Growth with 30 years of reality. Global Environmental Change (in press).

12. Hall, CAS & Day, JW. Revisiting the limits to growth after peak oil. *American Scientist* 97, 230–237 (2009).

13. IPCC. *Climate Change 2007: Synthesis Report, Summary for Policymakers* [online] (2007) (http://www.ipcc.ch).

14. Hirsch, R. Interview with Bob Hirsch: The stonewalling of peak oil [online] (2009) (http://peak-oil.org/2009/09/interview-with-bob-hirsch-the-stonewalling-of-peak-oil/).

15. Andrews, S. A few short clips from "The most recent economic downturn is a peak oil recession" [online] (2009) (http://peak-oil.org/2009/11/a-few-short-clips-from-the-most-recent-economic-downturn-is-a-peak-oil-recession/).

16. World oil production forecast: Update May 2009 [online] (2009) (http://www.theoildrum.com/node/5395).

17. Schumacher, EF. *Small Is Beautiful: A Study of Economics as if People Mattered* (Blond & Briggs, London, 1973).

18. Goldsmith, E, Allen, R, Allaby, M, Davoll, J & Lawrence, S. A blueprint for Survival. *Ecologist* 2, 1 (1972).

19. Hails, C et al. (eds.) *Living Planet Report 2008* [online], 15 (2008) (http://www.panda.org).

20. Schmidt-Bleek, F. *Factor 10 Manifesto* [online] (2000) (http://www.factor10-institute.org).

21. Pontin, J & Roderick, I. *Converging World: Connecting Communities in Global Change.* Schumacher Briefing No.13 (Green Books, UK 2008).

22. Russel, B. *In Praise of Idleness and Other Essays,* 15–17 (Allen and Unwin, London, 1935).

23. Mill, JS. *Principles of Political Economy* Revised edn, Vol. 2 (Colonial Press, New York, 1900).

24. Keynes, JM. *Essays in Persuasion* (MacMilland and Co., London, 1931).

25. Daly, H (ed). *Towards a Steady State Economy* (WH Freeman, San Francisco, 1973).

26. Costanza, R (ed). *Ecological Economics: The Science and Management of Sustainability* (Columbia UniversityPress, New York, 1991).

27. Jackson, T. *Prosperity without Growth? The Transition to a Sustainable Economy* [online] 6 (2009)(www.sd-commission.org.uk/publications/downloads/prosperity_without_growth_report.pdf)

28. Jackson, T. *Prosperity without Growth: Economics for a Finite Planet* (Earthscan, London, 2009).

29. Lahart, J, Barta, P & Batson, A. New limits to growth revive Malthusian fears. *Wall Street Journal* [online] (March 24, 2008) (http://online.wsj.com/public/article_print/SB120613138379155707.html).

30. Stiglitz, JE, Sen, A & Fitoussi, J-P. *Report by the Commission on the Measurement of Economic Performanceand Social Progress* [online], 12 (2009) (http://www.stiglitz-sen-fitoussi.fr/en/index.htm).

31. Rockström, J et al. A safe operating space for humanity. *Nature*, 461, 472, 24. September 2009)

32. Rockström, J et al. Planetary boundaries: Exploring the safe operating space for humanity. *Ecology and Society* [online] 14, 32 (2009) (http://www.ecologyandsociety.org/vol14/iss2/art32).

33. Nørgård, JS in *Energy Efficiency and Sustainable Consumption: The Rebound Effect* (eds Herring, H & Sorell, S) Ch. 10, Avoiding rebound through a steady state economy, 204–223 (Palgrave Macmillan, UK, 2009).

34. Lloyd, B. The Commons revisited: The tragedy continues. *Energy Policy* 35 5806-5818 (2007).

35. Hardin, G. Tragedy of the commons. *Science* 162 1243–1248 (1968).

36. The Sveriges Riksbank Prize in Economic Science in memory of Alfred Nobel [online] (2009) (http://nobelprize.org/nobel_prizes/economics/laureates/2009).

37. In praise of … Elinor Ostrom. *Guardian* [online] (October 13, 2009) (http://www.theguardian.com/business/2009/oct/13/nobel-prize-economics-elinor-ostrom).

38. Meadows, DH, Meadows, DL & Randers, J. *Beyond the Limits.* (Earthscan, London, 1992).

Section III

# VISION AND COMMUNITY ENGAGEMENT

# THE SCHOLARSHIP OF ENGAGEMENT

By Ernest L. Boyer

L et me begin with a self-evident observation: American higher education is, as Derek Bok once poetically described it, "a many-splendored creation." We have built in this country a truly remarkable network of research universities, regional campuses, liberal arts colleges, and community colleges, which have become, during the last half century, the envy of the world.

But it's also true that after years of explosive growth, America's colleges and universities are now suffering from a decline in public confidence and a nagging feeling that they are no longer at the vital center of the nation's work. Today the campuses in this country are not being called upon to win a global war or to build Quonset huts for returning GIs. They're not trying to beat the Soviets to the moon or to help implement the Great Society programs. It seems to me that for the first time in nearly half a century, institutions of higher learning are not collectively caught up in some urgent national endeavor.

Still, our outstanding universities and colleges remain, in my opinion, among the greatest sources of hope for intellectual and civic progress in this country. I'm convinced that for this hope to be fulfilled, the academy must become a more vigorous partner in the search for answers to our most pressing social, civic, economic, and moral problems—and must reaffirm its historic commitment to what I have chosen to call, this evening, the *scholarship of engagement*.

The truth is that for more than 350 years, higher learning and the larger purposes of American society have been inextricably interlocked. The goal of the colonial college was to prepare civic and religious leaders—a vision succinctly captured by John Eliot, who wrote in 1636, "If we nourish not learning, both church and commonwealth will sink." Following the revolution, the great patriot Dr. Benjamin Rush declared, in 1798, that the nation's colleges would be "nurseries of wise and good men, to adapt our modes of teaching to the peculiar form of our government." In 1824 Rensselaer Polytechnic Institute was founded in Troy, New York; RPI was, according to historian Frederick Rudolph, a constant reminder that America needed railroad builders, bridge builders, builders of all kinds. During the dark days of the Civil War, President Abraham Lincoln signed the historic Land Grant Act, which linked higher learning to the nation's agricultural, technological, and industrial revolutions. And when social critic Lincoln Steffens visited Madison in 1909, he observed that "in Wisconsin, the university is as close to the intelligent farmer as his pig-pen or his tool-house."

At the beginning of this century, David Starr Jordan, president of that brash new institution on the West Coast, Stanford, declared that the entire university movement in this country "is toward reality and practicality." Harvard's president, Charles Eliot, who was completing nearly forty years of tenure, said America's universities are filled with the democratic spirit of "serviceableness." And in 1896 Woodrow Wilson, then a forty-year-old Princeton University professor, insisted that the spirit of service will give a college a place in the public annals of the nation. "We dare not," he said, "keep aloof and closet ourselves while a nation comes to its maturity."

Frankly, I find it quite remarkable that just one hundred years ago, the words *practicality* and *reality* and *serviceability* were used by the nation's most distinguished academic leaders to describe the mission of higher learning—which was, to put it simply, the scholarship of engagement. During my own lifetime, Vannevar Bush of the Massachusetts Institute of Technology formally declared, while in Washington serving two presidents, that universities that helped win the war could also win the peace—a statement that led to the greatest federally funded research effort the world has ever known. I find it fascinating to recall that Bush cited radar and

penicillin to illustrate how science could be of practical service to the nation. The goals in the creation of the National Science Foundation—which led to the Department of Defense and the National Institutes of Health—were not abstract. The goals were rooted in practical reality and aimed toward useful ends.

In the 1940s the GI Bill brought eight million veterans back to campus, which sparked in this country a revolution of rising expectations. May I whisper that professors were not at the forefront urging the GI Bill; this initiative came from Congress. Many academics, in fact, questioned the wisdom of inviting GIs to campus; after all, these men hadn't passed the SATs—they'd simply gone off to war, and what did they know except survival? The story gets even grimmer. I read some years ago that the dean of admissions at one of the well-known institutions in the country opposed the GIs because, he argued, many of them would be married; they would bring baby carriages to campus, and even contaminate the young undergraduates with bad ideas at that pristine institution. I think he knew little about GIs and even less about the undergraduates at his own college.

But putting that resistance aside, the point is largely made that the universities joined in an absolutely spectacular experiment, in a cultural commitment to rising expectations—and what was for the GIs a privilege became for their children and grandchildren an absolute right. And there's no turning back.

Almost coincidentally, Secretary of State George C. Marshall, in 1947, at a commencement exercise at Harvard, announced a plan for European recovery, and the Marshall Plan sent scholars all around the world to promote social and economic progress. Ten years later, when the Soviets sent *Sputnik* rocketing into orbit, the nation's colleges and universities were called upon once again, this time to design better curricula for the nation's schools and to offer summer institutes for teachers. And one still stumbles onto the inspiration of that time. I remember having a lunch in Washington. We thought we were talking privately about the federal program to help teachers under the Eisenhower administration, only to find we were being overheard at the next table, which you should always assume in Washington. And the man stopped by and said, "I just wanted to tell you that I was one of the NDEA [National Defense Education Act] fellows at that time, and I've never had a better experience in my life." The inspiration of the teachers who came back from the summer institutes touched teachers all across the country. The federal government and higher education had joined with schools toward the renewal of public education.

Then, in the 1960s, almost every college and university in this country launched affirmative action programs to recruit historically bypassed students and to promote, belatedly, human justice.

I realize I've just dashed through three and a half centuries in three and a half minutes, more or less. What I failed to mention were the times when universities challenged the established order, when they acted appropriately both as conscience and social critic, and that, too, was in service to the nation. And there were other times when campuses were on the fringes of larger national endeavors, standing on the sidelines, failing to take advantage of opportunities that emerged. Still, I am left with two inescapable conclusions. First, it seems absolutely clear that this nation has, throughout the years, gained enormously from its vital network of higher learning institutions. At the same time it's also quite apparent that the confidence of the nation's campuses themselves has grown during those times when academics have been called upon to serve a larger purpose—to participate in the building of a more just society and to make the nation more civil and secure.

This leads me to say a word about the partnership today. To what extent has higher learning in the nation continued this collaboration, this commitment to the common good?

I hope I don't distort reality when I suggest that in recent years, the work of individual scholars as researchers has continued to be highly prized, and that also, in recent years, teaching has increasingly become more highly regarded, which of course is great cause for celebration. But it seems to me that it's also true that at far too many institutions of higher learning, the historic commitment to the scholarship of engagement, as I've chosen to call it, has dramatically declined.

I do a lot of work with colleges and universities, and study countless catalogs, and it won't surprise you to hear that almost every college catalog in this country still lists teaching, research, and service as the priorities of the professoriate. And yet it won't surprise you either that at tenure and promotion time, the harsh truth is that service is hardly mentioned. Even more disturbing, faculty who do spend time on so-called applied projects frequently jeopardize their careers.

Russell Jacoby, in a fascinating book entitled *The Last Intellectuals: American Culture in the Age of Academe,* observes that the influence of American academics has declined precisely because being an intellectual has come to mean being in the university and holding a faculty appointment (preferably a tenured one), writing in a certain style understood only by one's peers, and conforming to an academic rewards system that encourages disengagement and even penalizes professors whose work becomes useful to nonacademics—or *popularized,* as we like to say. Intellectual life, Jacoby said, has moved from the coffee shop to the cafeteria, with academics participating less vigorously in the broader public discourse.

But what I find most disturbing, as almost the mirror image of that description, is a growing feeling in this country that higher education is in fact part of the problem rather than the solution—going still further, that it's become a private benefit, not a public good. Increasingly, the campus is being viewed as a place where students get credentialed and faculty get tenured, while the overall work of the academy does not seem particularly relevant to the nation's most pressing civic, social, economic, and moral problems. Indeed, there follows from that the concept that if students are the beneficiaries and get credentialed, then let students pay the bill. And I've been almost startled to see that when the gap increases in the budget, it's the student, and the student fees, that are turned to automatically—after all, it's a private benefit, and let the consumer, as we like to say, pay the bill.

Not that long ago, it was generally assumed that higher education was an investment in the future of the nation—that the intellect of the nation was something too valuable to lose, and that we needed to invest in the future through the knowledge industry.

I often think about the time when I moved, almost overnight, from an academic post in Albany, New York, to a government post in Washington, DC. These were two completely separate worlds. At the university, looking back, I recall rarely having serious dialogues with "outsiders"—artists or "popular" authors or other intellectuals beyond the campus. And yet I was fascinated by Derek Bok's observation, on leaving his tenured post at Harvard, that the most consequential shifts in public policy in recent years have come not from academics but from such works as Rachel Carson's *Silent Spring*, Ralph Nader's *Unsafe at Any Speed*, Michael Harrington's *The Other America*, Betty Friedan's *The Feminine Mystique*—books that truly place environmental, industrial, economic, and gender issues squarely in a social context.

I teach occasionally at the Woodrow Wilson School of Public and International Affairs [Princeton University], and I open the first class by asking, "How is public policy shaped in America? Where does it originate? How does the debate get going?" Almost always, the undergraduates will start with the president, then Congress—or they might think of the state legislature. Then I ask them if they have ever heard of Rachel Carson or Michael Harrington, and a kind of bewildered look appears. And yet the truth is that out of the seminal insights of these intellectuals, public discourse begins—and very often Congress is the last, not the first, to act, trying to catch up with the shifting culture. So it is with the academy. One wonders why discourse between faculty and intellectuals working without campus affiliation can't take place within the academy itself.

But on the other hand, I left Albany and went to Washington, and I must tell you that I found government to be equally-or, to go one step further, even more startlingly—detached. In Washington we did consult, I want to assure you, with lawyers and political pressure groups, driven usually by legislative mandates and certainly by White House urges. But rarely were academics invited in to help put our policy decisions in historical or social or ethical perspective. Looking back, I recall that we talked for literally hundreds of hours about the procedural aspects of our work and the legal implications, but I do not recall one occasion when someone asked, "Should we be doing this in the first place?"—a question, I suspect, that could have been asked only by a detached participant with both courage and perspective.

Recently, I've become impressed by just how much this problem, which I would describe as impoverished cultural discourse, extends beyond government to mass communication, in which—perhaps with the exception of *The MacNeil/Lehrer News Hour* and *Bill Moyers*

*Journal*—the nation's most pressing social, economic, and civil issues are endlessly discussed primarily by politicians and self-proclaimed pundits, while university scholars rarely are invited to join the conversation.

Abundant evidence shows that both the civic and academic health of any culture is vitally enriched as scholars and practitioners speak and listen carefully to each other. In a brilliant study of creative communities throughout history, Princeton University sociologist Carl Schorske, a man I greatly admire, describes the Basel, Switzerland, of the nineteenth century as a truly vibrant place where civic and university life were inseparably intertwined. Schorske states that the primary function of the university in Basel was to foster what he called "civic culture," while the city of Basel assumed that one of its basic obligations was the advancement of learning. The university was engaged in civic advancement, the city was engaged in intellectual advancement, and the two were joined. I read recently that one of the most influential commentators achieved his fame not from published articles but from lectures he gave in the Basel open forum.

I recognize, of course, that "town" is not "gown." The university must vigorously protect its political and intellectual independence. Still, one does wonder what would happen if the university would extend itself more productively into the marketplace of ideas. I find it fascinating, for example, that the provocative Public Broadcasting Service program *Washington Week in Review* invites us to consider current events from the perspective of four or five distinguished journalists who, during the rest of the week, tend to talk only to themselves. I've wondered occasionally what *Washington Week in Review* would sound like if a historian, an astronomer, an economist, an artist, a theologian, and perhaps a physician, for example, were asked to comment. Would we be listening and

thinking about the same week, or would there be a different profile and perspective? How many different weeks were there that week? And who is interpreting them for America?

What are we to do about all of this? As a first step, and coming back to the academy itself, I'm convinced that the university has an obligation to broaden the scope of scholarship. In a recent Carnegie Foundation report entitled *Scholarship Reconsidered,* we propose a new paradigm of scholarship, one that assigns to the professoriate four essential, interlocking functions. We propose, first, the *scholarship of discovery,* insisting that universities, through research, simply must continue to push back the frontiers of human knowledge. No one, it seems to me, can even consider that issue contestable. And we argue, in our report, against shifting research inordinately to government institutes or even to the laboratories of corporations that could directly or indirectly diminish the free flow of ideas.

But while research is essential, we argue that it is not sufficient, and to avoid pedantry, we propose a second priority, called the *scholarship of integration.* There is, we say, an urgent need to place discoveries in a larger context and create more interdisciplinary conversations in what Michael Polanyi of the University of Chicago has called the "overlapping [academic] neighborhoods"—or in the new hyphenated disciplines, in which the energies of several different disciplines tend enthusiastically to converge. In fact, as Clifford Geertz of the Institute for Advanced Study has argued, we need a new formulation, a new paradigm of knowledge, since the new questions don't fit the old categories.

Speaking of bringing the disciplines together, several years ago, when physicist Victor Weisskopf was asked what gave him hope in troubled times, he replied, "Mozart and quantum mechanics." But where in our fragmented intellectual world do academics make connections such as these? We assume they live in separate worlds, yet they may be searching for the same interesting patterns and relationships, and finding solutions both intellectually compelling and aesthetic. I remember that during the days of the liftoffs at Cape Kennedy, when the rockets lifted successfully into orbit, the engineers wouldn't say, "Well, our formulas worked again"; they would say, almost in unison, "Beautiful." I always found it fascinating that they chose an aesthetic term to describe a technological achievement. But where do the two begin and end?

Beyond the scholarship of discovering knowledge and integrating knowledge, we propose in our report a third priority: the *scholarship of sharing knowledge*. Scholarship, we say, is a communal act. You never get tenured for research alone. You get tenured for research and publication, which means you have to teach somebody what you've learned. And academics must continue to communicate—not only with their peers but also with future scholars in the classroom—in order to keep the flame of scholarship alive. Yet the harsh truth is that on many campuses, it's much better to prepare a paper and present it to colleagues at the Hyatt in Chicago than to present it to the students right on campus, who perhaps have more future prospects than one's peers.

Finally, in *Scholarship Reconsidered,* we call not only for the scholarship of discovering knowledge, not only for the scholarship of integrating knowledge to avoid pedantry, not only for the sharing of knowledge to avoid discontinuity, but also for the application of knowledge to avoid irrelevance. And we hurriedly add that when we speak of applying knowledge, we do not mean "doing good," although that's important. Academics have their civic functions, which should be honored, but by *scholarship of application* we mean having professors become what Donald Schön of the Massachusetts Institute of Technology has called "reflective practitioners," moving from theory to practice,

and from practice back to theory, which in fact makes theory, then, more authentic-something we're learning in education and medicine, in law and architecture, and all the rest. Incidentally, by "making knowledge useful," we mean everything from building better bridges to building better lives, which involves not only the professional schools but the arts and sciences as well.

Philosophy and religion also are engaged in the usefulness of knowledge, as insights become the interior of one's life. Recently, I reread Jacob Bronowski's moving essay on science and human values, which, as you recall, was written after his visit in 1945 to the devastation of Hiroshima. In that provocative document, he suggests that no sharp boundaries can be drawn between knowledge and its uses. And he insists that the convenient labels of "pure" and "applied" research simply do not describe the way most scientists really work. To illustrate his point, Bronowski notes that Sir Isaac Newton studied astronomy precisely because navigating the sea was the preoccupation of the society in which he was born. Newton was, to put it simply, an engaged scholar. And Michael Faraday, Bronowski writes, sought to link electricity to magnetism because finding a new source of power was the preoccupation of his day. Faraday's scholarship was considered useful. The issue, then, Bronowski concludes, is not whether scholarship will be applied but whether the work of scholars will be directed toward humane ends.

This reminder that the work of the academy ultimately must be directed toward larger, more humane ends brings me to this conclusion: I'm convinced that in the century ahead, higher education in this country has an urgent obligation to become more vigorously engaged in the issues of our day, just as the land grant colleges helped farmers and technicians a century ago. And surely one of the most urgent issues we confront, perhaps the social crisis that is the most compelling, is the tragic plight of children.

In his first State of the Union Address, President George Bush declared that the nation's first education goal was that by the year 2000, all children in this country would come to school "ready to learn." Yet we have more children in poverty today than we did five years ago. Today a shocking percentage of the nation's nineteen million preschoolers are malnourished and educationally impoverished. Several years ago, when we at the Carnegie Foundation surveyed several thousand kindergarten teachers, we learned that thirty-five percent of the children who enrolled in school the year before were, according to the teachers, linguistically, emotionally, or physically deficient. One wonders how this nation can live comfortably with the fact that so many of our children are so shockingly impoverished.

These statistics may seem irrelevant in the hallowed halls of the academy or in the greater world of higher learning, yet education is a seamless web. If children do not have a good beginning—if they do not receive the nurture and support they need during the first years of life—it will be difficult, if not impossible, fully to compensate for the failure later on. My wife, a certified midwife, has convinced me that the effort· has to be made not only before school but surely before birth itself, during the time when nutrition becomes inextricably linked to later potential.

To start, higher education must conduct more research in child development and health care and nutrition. I do not diminish this role at all. This too is in service to the nation. But I wonder if universities also might take the lead in creating children's councils in the communities that surround them. The role of the universities would be to help coordinate the work of public and private agencies concerned with children, preparing annually, perhaps, what I've chosen to

call a "ready to learn" report card—a kind of environmental impact statement on the physical, social, and emotional conditions affecting children, accompanied by a cooperative plan of action that would bring academics and practitioners together. James Agee, one of my favorite twentieth-century American authors, wrote that with every child born, under no matter what circumstances, the potential of the human race is born again. And with such a remarkably rich array of intellectual resources, certainly the nation's universities, through research and the scholarship of engagement, can help make it possible for more children to be "ready to learn." Perhaps universities can even help create in this country a public love of children.

A second challenge, I'm convinced, is that colleges and universities must become more actively engaged with the nation's schools. We hear a lot of talk these days about how the schools have failed, and surely education must improve—but the longer the debate continues, the more I become convinced that it's not the schools that have failed; it's the partnership that has failed. Today our nation's schools are being called upon to do what homes and churches and communities have not been able to accomplish. And if they fail anywhere along the line, we condemn them for not meeting our high-minded expectations. Yet I've concluded that it's simply impossible to have an island of excellence in a sea of community indifference. After going to schools from coast to coast, I've also begun to wonder whether most school critics could survive one week in the classrooms they condemn. While commissioner of education, I visited an urban school with a leaky roof, broken test tubes, Bunsen burners that wouldn't work, textbooks ten years old, falling plaster, armed guards at the door—and we wonder why we're not world-class in math and science, or in anything, for that matter.

Especially troublesome is our lack of support for teachers. In the United States today, teachers spend an average of $400 of their own money each year, according to our surveys, to buy essential school supplies. They're expected to teach thirty-one hours every week, with virtually no time for preparation. The average kindergarten class size in this country is twenty-seven pupils, even though research reveals it should be seventeen. In one state the average kindergarten class size is forty-one. I've never taught kindergarten or first grade, but I do have several grandchildren, and when I take them to McDonald's or some other fast-food spot, I come home a basket case from just keeping mustard off the floor and tracking all the orders that keep changing every thirty seconds. And I'm not even trying to cram them for the SATs—I'm just trying to keep body and soul together. Class size does matter, especially in the early years, and it correlates directly with effective learning.

About a dozen years ago, the late Bart Giamatti invited me to evaluate what was called the Yale–New Haven Teacher's Institute. I was delighted to discover that some of Yale's most distinguished scholars directed summer seminars based on curricula that teachers themselves had planned. Incidentally, teachers in that program were called Yale Fellows—and, I was startled to discover, they were even given parking spaces on campus, which is about the highest status symbol a university can bestow. I'm suggesting that every college and university should view surrounding schools as partners; should give teaching scholarships to gifted high school students, just as we give athletic scholarships; and should offer summer institutes for teachers, who are, I'm convinced, the unsung heroes of the nation.

During my Yale visit, I dropped in on a sixth-grade classroom in New Haven. Thirty children were crowded around the teacher's desk, and I thought it was a physical attack—I almost ran to the central office for help. But then I paused and discovered they were there not out of anger but out of intense enthusiasm. They had just finished reading Charles Dickens' *Oliver Twist,* and they were vigorously debating whether little Oliver could survive in their own neighborhood-speaking of relating the great books and intellectual inquiry to the realities of life. The children concluded that while Oliver had made it in far-off London, he'd never make it in New Haven, a much tougher city. I was watching an inspired teacher at work, relating serious literature to the lives of urban youth today.

This leads me to say a word about higher education in the nation's cities. It's obvious that the problems of urban life are enormously complex; there are no simple solutions. Yet cities determine the future of this country—and many of our nation's children live in them. And I find it ironic that universities, which focused with such energy on rural America a century ago, have never focused with equal urgency on our cities. Many universities do have sponsored projects in urban areas—Detroit, Buffalo, New York, Philadelphia, and Baltimore, to name just a few. But typically, these so-called model programs limp along, supported with soft money. Especially troublesome is the fact that academics who participate are not professionally rewarded. Higher education cannot do it all, but Ira Harkavy of the University of Pennsylvania soberly warns that our great universities simply cannot afford to remain islands of affluence, self-importance, and horticultural beauty in seas of squalor, violence, and despair. With their schools of medicine, law, and education and their public policy programs, universities surely can help put our cities and perhaps—perhaps—even our nation back together.

Here, then, is my conclusion. At one level, the scholarship of engagement means connecting the rich resources of the university to our most pressing social, civic, and ethical problems, to our children, to our schools, to our teachers, and to our cities—just to name the ones I am personally in touch with most frequently; you could name others. Campuses would be viewed by both students and professors not as isolated islands but as staging grounds for action.

But at a deeper level I have this growing conviction that we need not just more programs but also a larger purpose, a larger sense of mission, a larger clarity of direction in the nation's life as we move toward century twenty-one. Increasingly, I'm convinced that ultimately, the scholarship of engagement means creating a special climate in which the academic and civic cultures communicate more continuously and more creatively with each other, helping to enlarge what anthropologist Clifford Geertz describes as the universe of human discourse and enriching the quality of life for all of us.

Many years ago Oscar Handlin put the challenge this way: "[A] troubled universe can no longer afford the luxury of pursuits confined to an ivory tower. ... [S]cholarship has to prove its worth not on its own terms, but by service to the nation and the world." This, in the end, is what the scholarship of engagement is all about—and, indeed, what the American Academy of Arts and Sciences provides precisely: a forum in which the nation can confront its mission in a larger, more enlightened sense.

# FROM SERVICE TO SOLIDARITY

## Engaged Education and Democratic Globalization

By Mark Wood

In *Scholarship Reconsidered: Priorities of the Professoriate,* the late Ernest L. Boyer and his colleagues provided a succinct history of higher education and made a persuasive argument for recovering and building on models of academic teaching and research that emphasize community engagement, public service, and civic responsibility.

> Now is the time to build bridges across disciplines, and connect the campus to the larger world. ... We need scholars who not only skillfully explore the frontiers of knowledge, but also integrate ideas, connect thought to action, and inspire students. ... If the nation's colleges and universities cannot help students see beyond themselves and better understand the interdependent nature of our world, each new generation's capacity to live responsibly will be dangerously diminished. *(Boyer 1990, 77)*

Since the publication of *Scholarship Reconsidered* in 1990 universities across the nation have taken up Boyer's provocative challenge to advance academia's democratic mission. Many institutions now invest as much time, energy, and resources in service-learning and community partnerships as they do multicultural curriculum and diversity recruitment. Numerous professional associations, including the Association for General and Liberal Studies and the American Association of Higher Education, now hold regular conferences concerned with developing curriculum that fosters community involvement, participatory democracy, and civic responsibility. A host of new journals concerned with addressing the pedagogical and institutional problems related to achieving these goals, including the *Journal of Public Outreach and Service* and the *Journal of Higher Education Outreach and Engagement,* have emerged over the past decade. In fact, the movement Boyer inspired grew so much during the 1990s that in 1999 presidents of fifty-one universities expressed their commitment to this movement in their *Presidents' Fourth of July Declaration on the Civic Responsibility of Higher Education.*[1] Emphasizing the link between education and citizenship, the presidents affirmed, "We must teach the skills and values of democracy, creating innumerable opportunities for our students to practice and reap the results of the real, hard work of citizenship. ..." *(Ehrlich 1999).*

With all these positive gains, however, the engaged education movement has been slowed by a variety of factors, not the least of which have been insufficient funding and inadequate recognition of the value of public teaching and scholarship for the purposes of promotion and tenure *(Peters et al. 2003, 85).* Drawing from the research of William Sullivan, a scholar at the Carnegie Foundation, Scott J. Peters notes that in addition to these factors, higher education operates "on a default program of 'instrumental individualism' that ignores explicit consideration of larger questions of social, political, and moral purpose" *(Peters 2004, 24).* By its nature this program impedes progress toward the university's teaching "the skills and values of democracy" and becoming, as Boyer contended that it should, "a more vigorous partner in the search for answers to our most pressing social, civic, economic, and moral problems" *(ibid; cited in Hill 2001/2002, 11).* To become such a partner in the grand democratic experiment requires, says Peters, that we continue to develop "a robust understanding of [higher education's] civic identity and mission" and, he adds, that we be careful "to infuse public service and outreach work with a civic rather than a market spirit." *(27).*

## PUBLIC SERVICE AND THE MARKET SPIRIT

Infusing public service and outreach work with a civic rather than a market spirit has indeed become a serious challenge to faculty across the country as market forces exercise increasing influence over the development of higher education. The business-trained administrators who increasingly manage educational institutions—what might more accurately be called education maintenance organizations[2]—are under growing pressure from state legislators, who are in turn under growing pressure from their corporate sponsors, to support curriculum and research that, as James Engell and Anthony Dangerfield write in "The Market-Model University," *(1998)* advances corporate agendas. In "The Kept University," Eyal Press and Jennifer Washburn note that "universities are behaving more and more like for-profit businesses," forming partnerships with corporations and investors, selling patents on technological and intellectual properties, marketing lectures and courses, and conceiving the mission of education as, on the one hand, fostering the growth of the capitalist relations of production and consumption and, on the other, providing students with the skills they need to secure employment *(2000, 39).*

The impact of the market-driven reorganization of higher education affects scholarship in a variety of ways. For example, when universities form partnerships with private investors, the latter frequently impose contractual obligations that, for example, restrict the freedom of scientists to share their findings (the life blood of scientific and technological development) and in some cases to publish their results. As Steven Rosenberg of the National Cancer Institute indicates, "the ethics of business and the ethics of science do not mix well" *(cited in Press and Washburn 2000, 42).* In addition to imposing restrictions on the exchange of information, corporations tend to fund research that is likely to result in the production of profitable commodities, even while research into non-profitable areas may address social and environmental problems that need to be addressed to ensure the universal well being of the human community *(see Silverstein 1999)*.

Corporate forces and market ideologies also influence curriculum development. Departments that do not prepare students for employment or support the expansion of capitalist relations of production and consumption have seen their funding reduced and/or been pressured to reform their curriculum to serve departments and programs that support these goals. The market-driven allocation of university resources is evident in the upsizing of programs with close ties to profit-making industries (e.g., marketing, engineering, and business), and downsizing of programs that do not hold the same promise (e.g., the humanities and social sciences). Between 1970 and 1994, note James Engell and Anthony Dangerfield, the number of students graduating with B.A.s in English, foreign languages, philosophy, and religious studies declined overall, even as the total number of B.A.s awarded to students increased. "Test what you will," write Engell and Dangerfield, "the humanities' vital signs are poor" *(1998)*.

Downsizing the humanities means that students graduate with at best a rudimentary knowledge of philosophy, religion, art, architecture, literature, and history. Moreover, to the extent that humanities courses offer students resources to develop their abilities to think critically (and thereby to investigate, analyze, and judge the validity of claims to truth), to make ethical judgments (and thereby to determine what is good, right, and just), to investigate the world and present their findings to others (and thereby to comprehend and educate fellow citizens), to appreciate beauty (and thereby to create environments that allow human beings

to flourish), and to imagine alternative realities (and thereby to envision more humane ways of living), and to the extent that developing these capacities is essential to living as responsible citizens, then downsizing the humanities means that our capacity for such a mode of living is being "diminished" *(Boyer 1990, 77)*.

Even as educational resources that enable students to investigate the world and imagine alternative possibilities for arranging our relations with each other and the earth are being downsized, students are being increasingly engineered to embrace capitalism's vision of the "good life." The presence of corporate retailers, vending machines, and human vendors selling their products on university campuses is now commonplace. Crossing campus recently I found myself weaving around Verizon trucks, tables, and employees hawking products in the student commons. Overhearing one student say, "What does Verizon have to do with higher education?" I found myself thinking, "Quite a lot," and imagined that the day we are required to wear corporate logos may not be far away. While it is true that higher education in the United States has regularly adjusted itself to satisfy the labor needs of business, academia increasingly supports the growth of capitalist production and consumption by educating students to be compliant workers and insatiable consumers.

In these ways, the market-model university compromises the democratic project of educating students to "see beyond themselves and better understand the interdependent nature of our world" and to live wisely as responsible global citizens *(Boyer 1990, 77; see Giroux 1998)*. Thus, precisely at a time when faculty are working to build an educational model that is "more sympathetically and productively involved with community concerns and needs," higher education is being dramatically restructured to advance the interests of corporate investors and the development of corporate-controlled globalization *(Ramaley 2003, 15)*.

## MARKET-MODEL EDUCATION FOR A MARKET-MODEL WORLD

The market model of education is problematic, however, not merely because it directs funding toward research that promises profits, downsizes courses that empower students to act as responsible citizens, and encourages obedience to the status quo. It is also because the market model supports a mode of social organization that daily proves unable to satisfy humanity's basic needs (e.g., food, water, housing, health care, education, and work), let alone ensure universal access to the resources individuals need to develop their abilities in a manner that is responsible to our shared conditions of social and natural existence.

In fact, the expansion of capitalist relations of production and the "liberalization" of these relations—that is, the expansion of what is often referred to as the freemarket and the weakening, if not outright abolition, of laws protecting workers and the environment[3]— have, according to the United Nations, UNICEF, and the World Bank, significantly widened the overall gap between the have-a-lots and have-too-littles and accelerated the pace of worldwide environmental despoliation.'[4] An "analysis of long-term trends in world income distribution (between countries) shows that the distance between the richest and poorest countries was about 3 to 1 in 1820, 11 to 1 in 1913, 35 to 1 in 1950, 44 to 1 in 1973 and 72 to 1 in 1992" *(United Nations 1999, 14)*. A recent study of global inequality and child poverty prepared for the United Nations Children's Fund (UNICEF) by the Townsend Centre for International Poverty Research at the University of Bristol contends that "the World Bank, IMF,

World Trade Organization and national governments—particularly of the G8 nations—have not only failed to reduce poverty but have exacerbated the problem" *(Frith 2003)*. Five hundred years of global capitalist development has, according to World Bank president James Wolfenson, resulted in a world in which "1.3 billion people live on less than one dollar a day; 3 billion live on under two dollars a day; 1.3 billion have no access to clean water; 3 billion have no access to sanitation; 2 billion have no access to electricity," conditions that could quite easily be improved if economic investment and human development were guided by the principles of human rights, social justice, and ecological sustainability *(cited in Randel, German, and Ewing 2000, 10)*.

Unfortunately they are not. Rather, and as is well known, investment and development under capitalism are guided primarily by the goals of profit maximization and capital accumulation. Unencumbered by religious values, moral imperatives, and humanitarian principles, corporations let millions of tons of grain rot in silos while tens of thousands of children die every day from starvation, *(Bread for the World Institute 2004)* invest billions of dollars into the production of horrifying weapons while billions of human beings live without education, health care, and housing *(see Udin 1996, Callari 2002, Fishman 2002, and Agence France-Press 2003)*, lobby against the production of generic AIDS medicines while AIDS engulfs the world in horror and suffering, and dump toxic waste into water supply systems while access to clean water is becoming as problematic as access to healthy food. In short, as long as profits are up, all is well in the kingdom of capital.[5]

It should also be noted that what is described as establishing "favorable business climates" often means supporting repressive governments and military interventions. Or, as Thomas Friedman *(1999)* wrote, "The hidden hand of the market will never work without a hidden fist—McDonald's cannot flourish without McDonnell Douglas, the designer of the F-15." As people around the world struggle to improve their conditions of life (e.g., by forming unions, demanding a larger slice of the pie they produce, and pressing for legislation to protect the environment and human rights), military and paramilitary forces frequently intervene to "secure the peace" and "establish democracy," that is, to make the world safe for multinational corporate investment and corporate-controlled development.[6] It is for this reason that Buddhist ethicist David R. Loy contends that "our global economy is institutionalized greed" and "our military-industrial complex is institutionalized aggression," and, following Friedman, we should note that the global economy and military-industrial complex generously support each other (1999, 86).

## RAISING QUESTIONS, RECOVERING HUMANITY

Lamentably, adds Loy, and as I've noted above, "our universities promote institutionalized ignorance of what is actually happening" (1999, 86). Most students learn little if anything about the history of colonialism, imperialism, the International Monetary Fund, the World Bank, the World Trade Organization, international debt, or for that matter, the nature of life for hundreds of millions of persons around the world, even as every aspect of their lives depends on these persons. They have never heard of, let alone studied, the United Nations Declaration on Human Rights or any of the other major international human rights documents and conventions. While a few have heard of the I.M.F., World Bank, and World Trade Organization, fewer have studied . how these transnational institutions operate, even as

these same nondemocratic institutions exercise growing influence over virtually every aspect of our lives. They frequently do not know the names of their elected representatives, let alone how to express their views to them. They are, at the most fundamental' level, frequently ill-prepared "to practice [let alone] ... reap the results of the real, hard work of citizenship ..." (Ehrlich 1999).

Following a recent presentation I gave in a senior-level international business course, a majority of students reported that none of the courses they had taken during their four or more years of college study had asked them to consider their ethical obligations with regard to such issues as poverty, health care, human rights, and the environment. Their ethical horizon had been reduced to their role as multinational corporate managers, or, more accurately, as several students forthrightly explained, to fostering the "bottom-line." Interestingly, many of these same students also expressed a feeling of being "ripped off." They wished they had been given the opportunity to reflect on what it means to be a human being, to analyze the potential social and environmental consequences of their actions as managers—in short, to "see beyond themselves" and to act as informed global citizens (Boyer 1990, 77).[7]

For all the rhetoric of accountability as a measure educational outcomes, given the current state of the world and continuing environmental and social trends, an educational model that does not prepare students to participate in the work of building a society that ensures more equitable conditions of development for all persons is not only unaccountable, it is irresponsible. In Solidarity and Suffering, Douglas Sturm writes:

> Education, at its best, poses the most fundamental questions we can ask: Who are we? What have we been? What might we become? How shall

we construct our lives, individually and collectively? What should be the shape of our tomorrow? ... At first blush, these questions appear—in some sense, they are—highly personal. ... But they are not merely personal. ... They are questions whose character compels us to consider the whole world in all its diversity and our place in it. Consciousness of *World* is a necessary correlate to consciousness of *Self*. *(1998, 221–22)*

Developing greater consciousness regarding the nature of the world and the self and the knowledge required to answer the questions posed above, begins with the recognition that, as described, the institutions that currently regulate our relations with each other and with nature are "unmoved by concern for the well-being of the earth's inhabitants and are driven by desire for their own profit and growth" *(Loy 2003, 100)*. That individual politicians, stockholders, corporate executive officers, and employees may care about poverty, the environment, and workers' rights does not alter in the least that the structure of corporate decision making and the obligations imposed by global competition for market share make it virtually impossible for CEOs or, for that matter, government administrations, to "be responsible in the ways that we need them to be" *(100)*. In practical terms this means "we cannot solve the [social and environmental] problems they keep creating by addressing the conduct of this or that particular corporation [or CEO], because the institution itself is the problem" *(101)*.

## TOWARD A RELATIONAL ECONOMY

Sturm suggests that solving our social and environmental problems is best accomplished by constructing institutions that are based on and supportive of a "relational vision" of human beings as "companions, each having a more or less direct influence on all others for better or for worse as we appropriate the resources of this world while shaping and directing the future" *(1998, 152)*. Achieving this goal most fundamentally requires subordinating productive property, what Marx referred to as the means of production, to "control by the community whose principal end is neither profit nor efficiency, but 'how to promote a better communal life'" *(88)*. Sturm adds, rightly in my opinion, that questions regarding how best to organize control and use of social and natural resources are simultaneously moral and spiritual questions, that is, questions regarding the nature and purpose of our individual and shared existence as human beings.

> From a broader, more theological perspective, the question of property takes on an appreciably different tone. In the final analysis, whose world is it anyway? Who properly controls the direction and disposition of the world? Who should benefit from the world's resources and how they are employed? Who should bear the burdens when all cannot benefit? Should not all those whose destiny is at stake in the shaping of the future be enabled to participate in that shaping? *(153)*

Constructing economic institutions that are subordinated to the will and responsible to the needs of the demos means constructing institutions that recognize and respect the dignity of human beings. It means creating an economy that does not reward behaviors that conflict with "the values of community, responsibility, virtue, stewardship, and a mutual concern for each other," as the current economic system does, but rather reinforces these values *(Fear and Sandmann 2001/2002, 31)*. At the same time, democratic ownership and control of the means of production "does not (necessarily) mean state possession, but it does mean public accountability (a question of control) and it entails concern for the public good (a question of use)" *(Sturm 1998, 153)*. The key aim of such institutions would be to "empower all those who have a stake in a productive process (workers, consumers, neighborhoods, and others, including, if only through surrogates, nonhuman creatures) to participate in the formulation of policies governing that process—what is produced, how it is produced, where it is produced, by whom it is produced" and, I might add, how what is produced is distributed *(153–154; see Smith 2002)*.

In light of the preceding analysis, I propose that our work as socially concerned scholars be oriented by the goals of democratizing the distribution of wealth and control of productive resources. Achieving these goals means constructing what Sturm calls "a relational economy" *(1998, 154)*. Such an economy would be organized explicitly to provide every individual with the resources he or she needs to develop his or her abilities in a manner that at the least does not undermine and at the most empowers other individuals to do the same. Such au economy would make it possible to realize what Sturm calls the elementary maxim of social justice: "So act that the life of the entire community and each of its participants might flourish" *(217)*. Working to construct a relational economy would empower our efforts to challenge conventional ways of doing business, advance critiques of the status quo, and take an ethical stand based on the shared conviction that humanity is best served by institutions that affirm the dignity of persons *(Fear and Sandmann 2001/2002, 36–37)*. Working to

build a relational economy means working to build institutions that are guided by and reward the values of community, equality, and democracy we strive to promote.

## FROM SERVICE TO SOLIDARITY

Like so many teachers inspired by Boyer's vision, I have sought to involve students not only in the classroom study of "our most pressing social, civic, economic, and moral problems" but also in the resolution of these problems through involvement with community organizations and projects *(cited in Hill 2001/2002, 11)*. I want to suggest, however, that advancing the goal of constructing a relational economy requires that we augment the ideas of service and good citizenship.

Outreach education frequently involves students serving the community by, for example, feeding homeless persons without necessarily challenging the institutions and social relations that divide society into those who own many mansions and those who lack a place to sleep at night. It means extending recycling programs without necessarily challenging the use of unsustainable forms of energy and an economic system that can exist only by promoting endless consumption. It means working to alleviate the suffering of HIV/AIDS patients without necessarily challenging the limits that insurance and pharmaceutical companies impose on access to health care and medicine.

This is not to say that feeding the homeless, extending recycling, and caring for HIV/AIDS patients are not valuable and vital projects. We can, however, and should augment the work of addressing existing needs and problems by engaging our students with projects that challenge the distribution and control of life-promoting resources. Borrowing from Suzanne C.

Toton, we may think about this as a shift from education engagement as *service* to education engagement as *solidarity (2002)*. Whereas the concept of service typically involves helping to heal the wounded (e.g., the homeless, hungry, and poor), the concept of solidarity involves working to address the conditions that wound individuals, communities, and nations in the first place. Whereas service maintains the distance between the server and the served, solidarity seeks to overcome this distance by creating generous conditions of life for all persons. Whereas service assumes the continuing existence of the present social order and on the basis of this assumption seeks to ameliorate human suffering, solidarity assumes the possibility of transforming the present social order and on the basis of this assumption seeks to build institutions that make it possible for all persons to flourish.

## FROM GOOD CITIZENS TO TRANSFORMED NON-CONFORMISTS

Moving from service to solidarity has implications for what it means to be a "good citizen." Teaching students how to act as good citizen's often means helping students develop the knowledge and skills they need to function within existing social, political, and economic institutions and relations. It assumes the existing system as natural and beyond need of serious challenge. Borrowing from Martin Luther King Jr.'s ideas concerning democratic citizenship, I propose that rather than help our students to become "good citizens," as this concept is most frequently understood, we ought to help our students become ethically, intellectually, and practically *maladjusted* to the forces of racism, sexism, militarism, environmental destruction, and class divisions that prevent human beings from enjoying dignified lives and keep the

human community locked in perpetual battle *(1981)*. We ought to educate our students to become what King describes as "transformed non-conformists." Doing so means helping our students to develop the intellectual, ethical, and creative capacities to participate in the work of building a society that enhances, rather than degrades, the dignity of human beings.

We can foster the development of transformed non-conformists through Classroom readings, lectures, and discussions that provide students with inspiring examples of persons who were, to borrow from Cornel West *(2000)*, freedom fighters for democracy, justice, and human rights, such as Sarah Grimke, Frederick Douglass, Sojourner Truth, Helen Keller, Dorothy Day, and Martin Luther King, Jr. And we can foster the development of transformed nonconformists by involving our students with community projects that seek to achieve a more democratic distribution of life-promoting resources.

Over the past three years many of my students have attended meetings and organized events in support of the Richmond Coalition for a Living Wage. This coalition of labor, religions, student, community, and neighborhood associations seeks to persuade the city council to pass a living wage ordinance that would require the city to pay a living wage to contracted and subcontracted employees. Raising awareness and obtaining community support are crucial to its passage. To this end students have organized campus rallies, voter registration campaigns, and benefit concerts, produced a documentary video on temp labor in the city, and developed a Web site (http://www.rclw.org/) for the coalition to make it easier to learn about and support living wage campaigns both here in Richmond and around the world. Combined with readings about Catholic, Buddhist, and Jewish perspectives on economic justice, some of which raised serious questions about the existing system of property relations, students not only learned about how the uneven distribution of money warps the practice of democracy *(see Plast 2003),* they began to raise questions about a system that allows a few to make tens of millions while tens of millions live on poverty-level wages.

The combination of studying global problems and participating in the living wage coalition has made it possible for students to appreciate the connections between poverty in Richmond and around the world, to build communities of solidarity, and to strengthen their ethical commitments. Students have created many resources, including a Web site, posters, and pamphlets, that community members now use to educate, organize, and demonstrate. Through their involvement with the movement students have been able to develop their organizational, pedagogical, and leadership skills and, just as important, they have discovered that getting involved provides a profound sense of meaning and purpose. Moreover, as students have shared their stories about the living wage movement with other students, roommates, and friends, many more have become involved. They begin to see how they can, given the interrelated nature of life, work in their own communities to change the world and that, as VCU student Charlie Schmidt explained, the world will improve just as soon as they "get busy" improving it. Getting busy is catching, it seems; so much so, in fact, that this past semester students brought the living wage campaign home to VCU. In this way, student involvement with the Richmond Coalition for a Living Wage has resulted in a vibrant "community of practice" *(Fear et. al. 2003, 59).*

## CONCLUSION

We can help students develop the theoretical and practical capacities to challenge all the conditions that prevent human beings from developing their particular abilities as individuals and that prevent them from participating fully in the creation of civilization. The consequences of helping our students develop these capacities are twofold. First, their effects help to improve our shared conditions of planetary life. Second, their effects enable each of them to live as a transformed non-conformist. As VCU student Archana Metha wrote, a "transformed non-conformist is a type of person who we should all strive to be like," adding that doing so not only helps to "improve the lifestyle of the people around us," it also enables students to "become better people. That is," she concludes, "the greatest lesson to teach a student" *(Metha 2003).* In this way, changing the self and changing the world creatively empower each other. As a result of participating in the wide-ranging work of building a humane global economy, students become hopeful that change is possible. They see through their own experience that we can achieve great things when we put our heads, hearts, and hands together. And this, after all, bodes well for us all.

## NOTES

1. By 2002, the number of presidential endorsors had grown to 459.
2. I borrow this idea from John O'Brien, art history professor at the University of British Columbia.
3. Among hundreds of trade disputes adjudicated by the World Trade Organization, not one judgment has favored workers, consumers, or nature. See http://www.globalexchange.org.
4. For information on global inequality see Global Issues at http:///www.globalissues.org. For information on the state of the world's ecology see the Union of Concerned Scientists at

http://www.ucsusa.org and the Environmental Defense fund at http://www.edf.org.

5. In addition of the almost $400 thousand million spent annually on the military, the U.S. approved almost $80 billion as a "down payment" for the war in Iraq. For the same amount of money, according to the National Priorities Project, the following could have been provided: 12,195,349 housing vouchers, 1,541,037 elementary school teachers, 361,253 fire trucks, 11,757,966 Head Start places for children, 34,830,861 children receiving health care, or some combination of these. See http://www.nationalpriorities.org for more on national spending.

6. Many religious organizations, including the Maryknoll Brothers and Sisters and the Buddhist monks of Nipponzan Myohoji, and secular organizations, including Amnesty International and Global Exchange, accuse the U.S. Army School of the Americas, at Fort Benning, Georgia, of training thousands of soldiers who, following their training, went on to participate in gross human rights violations, including disappearances, torture, and murder of unarmed men, women, and children, throughout Latin American nations. For more on the SOA see http://www.soaw.org.

7. It should be noted that several Business School professors at VCU are working to make ethical considerations a more central dimension of the Business School curriculum.

# REFERENCES

Agence France-Presse. 2003. U.S. world leader in arms sales, Saudi Arabia number 1 buyer. http://www.truthout.org/docs_03/101903F.shtml (October 2003).

Boyer, E. T. 1990. *Scholarship reconsidered: Priorities for the professoriate.* San Francisco: Jossey-Bass.

Bread for the World Institute. 2004. http://www.bread.org/institute/.

Brecher, J., and T. Costello. 1998. *Global village or global pillage.* 2d edition. Cambridge, Mass.: South End Press.

Callari, R. 2002. "The Enron-Cheney-Taliban connection?" *Albion Monitor* 28 February.

Ehrlich, Thomas. 1999. Presidents' declaration on the civic responsibility of higher education. http://www.compact.org/presidential/plc/plc-declaration.html. (August 2003).

Engell, J., and A. Dangerfield. 1998. The market-model university: Humanities in the age of money. *Harvard Magazine,* May–June. http://www.harvard-magazine.com/issues/mj98/forum.html (April 2003).

Fear, A. F., K. Bruns, L. Sandmeyer, A. M. Fields, S. Buhler, B. Bumham, and G. Imig. 2003. Experiencing engagement: stories from the field. *Journal of higher Education Outreach and Engagement* 8(1): 59–74.

Fear, F. A., and L. R. Sandmann. 2001/2002. The "new scholarship": Implications for engagement and extension. *Journal of higher Education Outreach and Engagement* 7(1&2): 29–39.

Fislunan, T. C. 2002. Making a killing: The myth of capital's good intentions. *Harper's,* August, 33–41.

Friedman, T. L. 1999. A manifesto for the fast world. *New York Times Magazine,* March 28. Available at Worldwide Media Relations http:// www.west.net/wwmr/fasl:wrld.htm (March 2003).

Frith, Maxine. 2003. Global trade keeps a billion children in poverty, says Unicef. *Independent Digital,* 22 October. http://news.independent.co.uk/world/politics/story.jsp?story=455868 (October 2003).

Giroux, H. 1998. The business of education. *Z Magazine,* July/August. http://www.zmag.org/ZMag/articles/girouxjulyaug98.htm (April 2003).

Greider, W. 1997. *One world, ready or not.* New York: Touchstone.

Hill, M. 2001/2002. Reflections of a new outreach journal editor. *Journal of higher Education Outreach and Engagement* 7(1&2): 9–14.

King, M. L., Jr. 1981. *Strength to love.* 1963 reprint. Philadelphia: Fortress Press.

King, M. L., Jr. 1986. *A testament of hope.* Edited by James M. Washington. San Francisco: Harper San Francisco.

Loy, D. 1999. The spiritual roots of modernity: Buddhist reflections on the idolatry of the nation state, corporate capitalism and mechanistic science. In *Socially engaged Buddhism for the new millennium.* Edited by Sulak Sivaraksa, 86–113. Bangkok: Sathirakoses-Nagapradipa Foundation.

Loy, D. 2003. *The great awakening: A Buddhist social theory.* Somerville, Mass.: Wisdom Publications.

Metha, A. 2003. Entry from journal for *Global Ethics and the Worlds Religions.*

Peters, S. 2000. The formative politics of outreach scholarship. *Journal of Higher Education Outreach and Engagement* 6(1): 23–30.

Peters, S., N. Jordan, T. Alter, and J. Bridger. 2003. The craft of public scholarship in land-grant education. *Journal of higher Education Outreach and Engagement* 8(1): 75–86.

Palast, Greg. 2002. The globalizer who came in from the cold. *Alternet, 19* March. http//www.alternet. org/story.html?StoryID=126527 (11 December 2003).

Palast, Greg. 2003. *The best democracy money can buy.* Rev. American edition. New York: Plume.

Press, E., and J. Washbum. 2000. The kept university. *Atlantic Monthly,* March, 39–54.

Randel, Judith, Tony German, and Deborah Ewing, eds. 2000. *The reality of aid 2000.* London: Earthscan.

Silverstein, Ken. 1999. Millions for Viagra, pennies for diseases of the poor. *The Nation, 19* July. http:// www.thenation.com/doc.mhtml?i=19990719&s =silverstein.

Smith, J. W. 2002. *Economic democracy: The political struggle of the twenty-first century.* 2d edition. Bloomington, Ind.: 1st Books Library.

Stiglitz, J. E. 2002. *Globalization and its discontents.* New York: W. W. Norton.

Sturm, D. 1998. *Solidarity and solidarity: Toward a politics of relationality.* Albany, N.Y.: SUNY Press.

Toton, S. C. 2002. Liberating justice education: From service to solidarity. *Journal for Peace & Justice Studies* 12(2): 231–48.

Townsend, Peter, and David Gordon, eds. 2003. *World poverty: New policies to defeat an old enemy.* Bristol: Policy Press.

Udin, Jeffrey. 1996. The Profits of Genocide. *Z Magazine, May.* http://www.zmag.org/ZMag/ articles/may96udin.htm (June 2003).

United Nations. 1999. Human Development Report 1999. http://hdr.undp.org/reports/globaV1999/ en/pdf/hdr_1999_full.pdf (11 December 2003).

West, C. 2000. *The Cornel West reader.* New York: Basic Civitas Books. See especially: 3–33, 149–73, and 425–34.

Section IV

# ETHICS, LAW AND POLICY

# IS THERE AN ETHICAL OBLIGATION TO ACT SUSTAINABLY? THEORIES OF ETHICS

By Tom Russ

The most erroneous stories are those we think we know best—and therefore never scrutinize or question.

Stephen Jay Gould

Up to now ethics have been humanist in purpose and character. We have reached a point where our casual impact and activities have profound effects on the environment and where the gap between the First and Third Worlds is growing. Thomas Berry correctly observed that the argument is usually framed as the good of nature against the bad of society or as the bad of society against the good of society. Berry suggests that each side has good and bad points and that we should compare the good with the good and the bad with the bad.[1] We need to answer the question of whether ethics stretches the cultural limits further to include eccentric elements—respect for life diversity and community.

Normative theories of ethics are about principles for distinguishing right from wrong. In general, normative ethical theories can be distinguished as either consequentalist or nonconsequentalist. Consequentalists believe that the moral rightness of an action is determined by its outcome: If the outcome is good, then the act is right and moral. Nonconsequentalist (deontological) theories contend that right and wrong are determined by more than the consequences of an action. Consequences may be important, but nonconsequentalists believe that other factors are also relevant to moral assessment. Professional ethics tend to be consequentalist in nature. In other words, it is the outcome of the design that is weighed, not the underlying rules or motivations for the design.

Questions about consequentalist ethics usually deal with concerns about who the outcome is good for. Should consequences be considered for all or for just the actor? This is particularly important when we consider the negative

Note: this chapter has been excerpted from the original publication.

obligations included in some professional codes of ethics that require the professional to prevent harm. Who in the array of stakeholders in a given design is in a position to define harm? Of course designers working in a marketplace must provide positive outcomes beyond their own or they will not be working for very long; the work of designers has multiple stakeholders. The stakeholders include the client, those who build or manufacture the product of design, those who distribute and sell the product, and the end user. The stakeholder shortlist is frequently expanded to include the marketplace, the designer's own profession, and the interests of the public at large as well. So design is evaluated by how well the designer's work meets the expectation and the needs of the various stakeholders. How well the designer satisfies this need determines the design to be good and an ethical act; to the degree it fails to satisfy, it is not good and not ethical from a professional sense. So from a consequentialist view, the end product is an ethical statement of sorts.

Moral rules are fundamentally derived from the obligation to respect every human, but moral obligations may be derived from many sources. Consider, for example, WD Ross's prima facie duties: A prima facie obligation is an obligation unless it is overridden by more pressing obligations.[2] McMahon goes further to assert an argument for moral intuition, a "spontaneous moral judgment."[3] Such intuition enables a person to make a moral judgment in circumstances not previously encountered. Intuition, on the other hand, is likely to be guided by our learned moral framework rather than be a wholly new moral paradigm. Moral rules are not static. Our judgment reflects our experience and our interests. As conditions change and as we learn as individuals and then as cultures, our moral standards change to reflect our new circumstance. In the past, slavery, institutionalized class systems, and cruelty to animals were accepted as moral, but over time the moral barometer changed as the interests of the society changed. This expanding moral umbrella is a function of the security and stability of the human species and cultures. The scope of inclusion expands to cover more people and living things as our fates intermingle. Our moral vision includes others, at least in part, as a function of our ability to meet our own needs first (i.e., Abraham Maslow's hierarchy of needs). In the absence of those needs, our social scope of moral inclusivity shrinks. Likewise, as our needs are met and secured, the aperture of moral standards opens.

Morality, it seems, changes then to reflect our awareness of an objective thing or circumstance and an understanding of how we relate to it. Our concern is lately expanding to include our environment.[4] If, however, morality is based on a cogent evaluation of our behavior it is reasonable to assume that some knowledge, appreciation, or awareness of the environment is required.

## EVALUATING ETHICAL OBLIGATIONS

In general, ethics can be evaluated in three ways: in terms of the Kantian categorical imperative, in terms of virtue morality, or in terms of utility.

### Categorical Imperative

Emmanuel Kant (1724–1804) posited that we should always act in such a way that we can will the maxim (a principle of action) of our action into a universal law. A thing is morally right if and only if we can "will it to become a universal law of conduct." A universal law in Kant's view is understandable, can be discovered through reason by anyone, and would apply to all people all of the time. Kant believed that the views of morality held by most people are pretty much correct and that all rational beings will rightfully pursue their own happiness. According to Kant, when people act for a reason—that is, when they decide to act—they are following a maxim, or a rule of action. He did not believe that each of us establishes a rule of action for each decision but that we can discover our reasons for acting if we stop and think about it. A maxim takes the form of: I will do action X in circumstance C for purposes P. Kant tested maxims by looking for contradictions in their universality. Inconsistency can arise in a maxim as a universal law in two ways: (1) if, as a universal law, it would jeopardize our survival or happiness; and (2) if we will something to be a universal law that cannot be a universal law. Kant believed that we have a duty to do the moral thing and that the decision to do the moral thing must be based on the duty and not on a presupposition of a favorable outcome or some other motivation; doing the right thing is the right thing to do. He crafted the categorical imperative to express this: "Act so that you treat humanity, whether in your own person or in that of another, always as an end and never as a means only."[5]

The categorical imperative is an unconditional directive: Do this, or don't do that. What makes an action right is that the agent would be willing to be so treated if the positions of the parties were reversed and that humans beings are treated as ends in themselves rather than means—in short, the familiar Golden Rule. Several duties are derived from the categorical imperative: (1) Everyone has a duty to help those in need; (2) everyone has an obligation to express gratitude because of a benefit received; and (3) everyone needs to possess a general respect for other people.

A more recent distillation of these ideas is found in the work of John Rawls (1921–2002), who was concerned about justice as it is expressed in our actions. Rawls wrote that the correct standards of justice are those that rational, self-interested individuals in a certain hypothetical state analogous to the state of nature (i.e., the original position) would accept if given the choice. Key to Rawls's theory is that the individual does not have to accept the standards; it is enough only that he or she would accept them if the choice existed. We can extend Rawls's ideas to moral and ethical choices. In this view it is important that the only acceptable moral rules or ethical acts are those everyone can agree on and that can be publicly justified. The hypothetical process Rawls proposes is what makes his approach interesting; he required that there be a "veil of ignorance" drawn over the members of society; that is, as they describe these rules of moral and ethical behavior they can have no personal information of themselves. All decisions about principles are made behind a veil of ignorance that assures impartiality and objectivity. Rawls said that all rational, self-interested individuals in the original position acting behind a veil of ignorance would agree on two principles to regulate the distribution of freedoms and economic resources in society:

1. Each person is to have an equal right to the most extensive basic liberty compatible with a similar liberty for others.
2. Social and economic inequalities are to be arranged so that they are both (a) the greatest benefit of the least advantaged and (b) attached to offices and positions open to all under conditions of fair equality of opportunity.

Moral rules and ethics are derived from these obligations to respect every human.[6]

## Virtue Ethics

Virtue ethics are derived from Greek philosophy wherein the primary function of morality is to cultivate virtuous traits in a person and wherein virtues are acquired through practice. Like Kant believed, the acquisition of virtues and moral behavior depend on appropriate motivation. If one is raised to be virtuous, one will act in ways consistent with those virtues. Aristotle believed that it was the obligation of the state to educate its citizens in the various virtues because it had an overarching interest in having virtuous citizens. A virtuous person acts on the learned virtues as his or her own reward. A virtuous person acts in an ethical manner because to do otherwise would be inconsistent with the virtues he or she has embraced and learned.

## Utilitarianism

Utilitarian ethics are concerned with producing the greatest possible good or happiness for everyone. Since utilitarians evaluate moral goodness based on the consequences, almost anything may be morally right or wrong depending on the circumstances. Utilitarians also wish to maximize happiness over the long run so that some early reduction in happiness may be acceptable if there is the promise of greater future happiness. Since we cannot know the future with certainty, we must act on what is likely to produce the best outcome. Although we are concerned with the total happiness we do not have to ignore our own happiness in the process.

Utilitarianism is concerned with outcomes, but when considering the greatest happiness we must also consider unhappiness. Utilitarians are often categorized as either act or rule utilitarians. Act utilitarianism considers an act right if and only if the good or benefit of the act outweighs the bad of the act. Good and bad are determined not solely in terms of the actor but also for the benefit or harm of all. It requires the moral agent to think of the consequences for all. Rule utilitarianism, on the other hand, distinguishes between an act and a rule. Rule utilitarianism requires that moral rules be observed and that these rules be followed based on the greatest utility or tendency to promote happiness.

Jeremy Bentham (1748–1832) used the concept of utility to describe the tendency of an act or a thing to increase or decrease happiness. Morality requires an actor to pursue the common good rather than his or her own happiness exclusively, but the notion of a common good is vague. Good and bad differ from person to person. Bentham tried to refine the notion of a common good. He wrote that the community does not exist as a body of members but, rather, that the community exists as the "sum of the interests" of the members. In his view, then, a thing is good if it advances or promotes the interests and is bad if it inhibits or constrains the interests. Among possible actions, any action that increases the interests or happiness of the whole would be a moral choice.[7]

John Stuart Mill (1806–1873) argued that rightness and wrongness are relative and that they occur as matter of degree: An act is right in proportion to its tendency to promote happiness. The more happiness it will produce, the more right it is. This is known as Mill's greatest happiness rule. The standard, though, is not the agent's own happiness but the greatest amount of happiness taken altogether. Mill suggested that we should follow moral rules and consider the utility of a choice only when there is a conflict between rules.[7]

Problems arise if act utilitarianism might require a person to forego his or her own interests or happiness in favor of increasing the total happiness. Bentham suggested a negative formulation of the greatest happiness principle, requiring instead that an action is wrong if and only if it would reduce total happiness and if alternatives can be found that would not reduce total happiness. Act utilitarianism ignores the distribution of happiness; as long as the greatest utility or happiness is achieved, the distribution is unimportant.

Critics of utilitarianism question whether it is workable. Is it possible to even know enough to be able to determine the greatest happiness or unhappiness? Further, can something be wrong even if it produces the greatest good? Utilitarians might argue that the greatest happiness for most people is moral and ethical regardless of great poverty or deprivation of a few. Critics find utilitarianism unjust and disagree that the distribution of happiness is a fair measure of moral or ethical behavior.

## Professional Practice Standards

Professional practice standards include the obligations and other standards of ethical conduct by professionals. The acceptable conduct is usually referred to as standards of "due care" or as standard of care and is determined by the customary practices of the professional community. Those without the requisite expertise are unqualified to determine what the standard should be because they do not have the special knowledge that distinguishes the professional. Some of the difficulties with relying on professional standard of care as a guide to behavior is that the practice standard may not be suited for some circumstances where not enough information is known or the information relied on is inaccurate. There is a presumption in the standard of care model that professionals actually have sufficient knowledge and expertise to determine necessary precautions or standards of performance. It is equally likely that evaluations of due care are based not solely on objective professional judgments of facts but on ethical ones as well. As knowledge moves forward on many fronts, how do the professions assure themselves that the standard of care reflects what is actually believed to be true in the natural world and in society? How might a change in the moral reasoning of society lead to changes in the standard of professional care? The standard of care would change to reflect this newer reasoning of course.

## What Obligations Do We Have to Other Living Things?

... Although there are fundamental values common to all cultures, the particulars of moral behavior are a function of the values and circumstances of a given society at a given time and place. In this way, a belief in the sanctity of life, for example, might be expressed in different ways and at different times. Morally reasoned rights originated primarily in special relationships between people (e.g., craftsman–customer, teacher–student, parent–child, neighbors), but a wider appreciation of the common interests of all people has led to wide recognition of human rights that are not simply functions of relationships.

Human rights are universal (i.e., everyone, everywhere, and at any time has these rights), are equal (i.e., no one person's rights are greater or less than another's), are not transferable, cannot be relinquished, and are natural rights (i.e., not derived from human institutions).[8] Negative human rights reflect the interests human beings have from being free from outside interference. Positive human rights are the interest we have in receiving certain benefits (e.g., right to education, right to life, political participation). Any right a person has also a correlative duty to act in a certain way. For example, you have a right to be fairly treated, but in turn you must treat others fairly. These rights can be expressed as negative rights (a person is enjoined from certain behaviors) and positive rights (a person has a duty to do something or behave in a certain way).

Until recently little thought was given to extending moral standing to other living things or to the future. For the most part, such attention that was given viewed other living things primarily in terms of how they might best be used toward the interests of people. For example, in the first treatise of government, John Locke justified the use and inferiority of plants and animals in the book of Genesis and went on for several pages listing the various rankings and uses.[9] In his lecture "Duties Toward Animals and Spirits" Kant said, "Our duties toward animals are merely indirect duties towards humanity. … If then any acts of animals are analogous to human acts and spring from the same principles, we have duties towards the animals because this we cultivate the corresponding duties towards human beings" and "Tender feelings towards dumb animals develop humane feelings towards mankind."[10]

In the 17th century, philosophers assessed the world in terms of primary and secondary qualities. Primary qualities consisted of the physical aspects of an object: its size, weight, material, how it moved, and similar characteristics that embodied the thing. Secondary qualities were those things that existed because of the interaction of the object and the observer. Characteristics such as color, taste, feel, and smell were functions of the interaction between observer and object; consequently, these secondary qualities were believed not to be entirely of the object itself. The role of science, then, was to describe the primary qualities of a thing. Secondary characteristics were not the purview of science.[11] In the intervening years, as science provided a clearer understanding the lines between primary and secondary qualities blurred so that today such distinctions are not routinely thought to exist.

The fundamental distinction between objectivity and subjectivity in Western philosophy is an important consideration in understanding the moral reasoning that supports an environmental ethic. Science studies the objective world, the real world, and is said to be "true," whereas the common interpretations of this world by individuals are subjective. The subjective world is one formed from perception and weighted by value and self-interest; subjective findings are not "true" in the sense of objective findings. Our subjective views, though, have been routinely challenged as our understanding of the objective world improves. Moral standards are concerned with behavior that is of serious consequence to human welfare and accordingly that are significantly informed by our subjective views of the world. Ultimately, however, the soundness of a moral standard depends on the adequacy of the underlying reasons and hence is powerfully influenced by science.

## UTILITARIAN VIEWS OF NATURE

The view of Kant and other early philosophers that other living things are subject to the interests of humans continues today. In his piece "People or Penguins," William F. Baxter presented a modern, utilitarian view of nature and the environment.[12] Baxter asserted that every environmental issue should be decided in favor of the greatest satisfaction of human interests based on four criteria[13]:

1. To discuss a problem it is necessary to be able to describe the problem in objective terms.
2. There should be no waste: In his words, nothing should be "employed so as to yield less than [it] might yield in human satisfaction."
3. Every human should be regarded as an end not mean to the betterment of another person.

4. Every person should have an equal opportunity to improve his or her share of satisfactions.

The basis for this formulation is articulated in six points that are in essence a list of the difficulties in objectifying other living things or intrinsic environmental values and in ascribing human values to nonhuman objects[14]:

1. People will always act in their own interests because it is their nature and will not act in another way.
2. Nature will be preserved because humans need it to be preserved. There is no "massive destruction of nonhuman flora and fauna" threatened.
3. What is good for humans is often good for "penguins and pine trees," so in this way humans are "surrogates for plant and animal life."
4. There is uncertainty as to how any other system could be administered.
5. If nonhuman things are to be valued as ends rather than means, it requires that someone determine how much each one counts and how he or she is to express his or her preferences.
6. These questions raise the question of what we ought to do, which in the end is uniquely human and meaningless to nonhumans.

Instead, Baxter argued that all of our actions should be assessed only in terms of his four criteria and the maximization of human satisfactions. His argument, however, contains several flaws. First, he required that nonhuman subjects have an objectively assessed value as a precedent to moral consideration but presumably could provide no such objective value for a human life. How would we value the human life? On the basis of the economic value of its chemical components or on the income earned or potentially earned in a life at work? What about the value of contributions to society of a great teacher, leader, or artist?

Second, Baxter assumed that the environment and the economy are a zero sum game, that a unit of work or resource expended there by definition requires an offsetting shortage somewhere else. He supposed that the costs associated with pollution control are necessarily expressed in "terms of other goods we will have to give up to do the job."[15] In fact, pollution controls create jobs and offset the externalities of no pollution control. Fairly robust economies have thrived without apparent offsetting negatives affects of pollution controls, as Baxter suggested.

Third, Baxter assumed that some central authority balances this real world. In fact, the world and our relationship to nature are far more complex than his argument recognizes. Science has found objective links between human interests and the environment that he does not account for. Fourth, he suggested that the only workable relationship with nature is one of subservience because of human nature. In fact, this is a problem of modern human society. Many other cultures do recognize our connectedness to nature and duty to it. These cultures are successful and also human but use a different worldview from the one Baxter assumed to be necessary. Last, Baxter's view of satisfaction does not account for human satisfaction in the future and assumes that the satisfaction of current generations is our only duty. In the end, the utilitarian view of our obligations to other living things, at least as argued by Baxter, is insufficient. We know, for example, that living and nonliving parts of our environment frequently have values that far outweigh economic or short-term human preferences. We understand that the environment is systemic and functions because of the interaction of many living and nonliving parts of which Baxter's approach simply does not account.

## SPECIESISM

By the 1970s our understanding of the inter-relatedness of living things and the systemic fabric of life prompted some to begin to call for a greater moral inclusion and the recognition of the moral value of other living things, if not actual moral rights.[16] Peter Singer's premise begins with a "fundamental presupposition" of the "basic moral principle" that all interests are entitled to equal consideration—that is, that any being with moral standing counts, or matters. Singer writes, "The argument for extending the principle of equality beyond our own species is simple, so simple that it amounts to no more than a clear understanding of the nature of the principle of equal consideration of interests. ... Our concern for others ought not to depend on what they are like, or what abilities they possess. ..."[17] In response, Joseph Des Jardins appropriately asks, "What characteristic quali-fies a being for equal moral standing?"[18] Singer quotes Jeremy Bentham:

> The day may come when the rest of the animal creation may acquire those rights which never could have been withholden from them but by the hand of tyranny. The French have already discovered that the blackness of the skin is no reason why a human being should be abandoned without redress to the caprice of a tormentor. It may one day come to be recog-nized that the number of legs, the villosity of the skin, or the termination of the os sacrum, are reasons equally insufficient for abandoning a sensitive being to the same fate. What else is

it that should trace the insuperable line? Is it the faculty of reason, or perhaps the faculty of discourse? But a full grown horse or dog is beyond comparison a more rational, as well as more conversable animal, than an infant of a day or a week, or even a month, old. But suppose they were otherwise, what would it avail? The question is not, Can they reason? Nor can they speak but, *Can they suffer?*[19]

Singer then says, "If a being suffers, there can be no moral justification for refusing to take that suffering into consideration."[20] "If we make a distinction between animals and [severely intellectually disabled adults, infant orphans], how do we do it, other than on a basis of a morally indefensible preference for members of our own species?"[21] In the section "Speciesism in Practice," Singer goes on to discuss animals as food. He observes that people in environments in which animals must be killed for food are justified, but the rest of us who are not in such straits cannot morally justify taking animals for food. The cruelty of the treatment of animals in the conversion of calories, treating animals "like machines" is specifically enjoined in Singer's book. It is interesting, though, that he seems to be less concerned about free-range and meadow-raised livestock than about the factory farm. While I agree that factory-farming prac-tices are cruel and deserve our condemnation, I am not sure it is an argument for extending "rights" to livestock. Can Singer argue that a free-range chicken is being treated better than a chicken raised in a wire cage as it is harvested? Were the evils of racism and sexism mitigated simply by eliminating slavery or chauvinism?

*Speciesism* is a term used by Singer as a cor-ollary to sexism and racism.[22] He supposes that it is wrong to deny other species moral standing on the basis of species. Our anthropocentric bias, according to Singer, can be likened to race bias. Racial prejudice, however, is intraspecies prejudice rather than interspecies prejudice. Bias in the form of preference is morally permissible. It would be morally acceptable, even expected, that, all other things being equal, I would act to save my child, my sibling, my friend, my neighbor, my countryman in a sort of descending order of preference—in order of my bias. Clearly, all other things being equal, there should be an expected bias in favor of and in preference for our own species.

Ultimately, Singer must deal with what Baxter referred to as the administration of the system he proposes. How would such a world work? How would the interests of one species be balanced against the interests of another? Whose interests and rights would have greater weight, and what would the basis of those distinctions be? In the end, predator eats prey. Human beings are in effect managing the world consciously or unconsciously. We have become so pervasive in our influence and effect that we must take our behavior into account. Morals are fundamentally based on human rather than on individual interests. Morals are different from etiquette and law. An action may be legal but unethical or immoral, or an action may be illegal but morally right. The law can be seen as the floor of moral conduct; that is, the minimum acceptable level of morality and understanding of the origins of moral standards may be less important than whether they can be justified.

Most philosophers agree that waste is wrong. Baxter used it as one of his four criteria. Singer's argument is at least partially rooted in the waste involved in our treatment of animals as food. Even Locke limited property to what could be held without spoilage and enjoins against waste: *"God has given us all things richly [sic]* is the Voice of Reason confirmed by Inspiration. But how far

has he given it us? *To enjoy.* As much as any one can make use of to any advantage of life before it spoils; so much he may by his labour fix a Property in. Whatever is beyond this is more than his share and belongs to others. Nothing was made by God for man to spoil or destroy."[23]

We have begun to realize that we are indeed subject to our environment and the rules of nature while at the same time we are affecting the systems that compose nature. As moral animals we must take responsibility for what we do. As intelligent animals we must reflect on the objective world science reveals to us and incorporate it into our subjective understanding. Further, perhaps the argument is not whether individual components of nature have rights but whether nature as a system does. If the absence of sentience is a concern, perhaps we can simply extend our moral umbrella to cover future generations of our species that will require resources and environmental quality composed of diverse functioning ecosystems. Our moral duty is not to a single species but to the protection of nature itself. A single species may be singled out and made more or less valuable in isolation from the ecosystem in which it lives naturally. The value to the system of nature may be incalculable because we simply do not understand how this species interacts in its ecosystem in the long or short term.

What obligations do we have to other living things? In the end, moral consideration of other living things appears to be in order. Perhaps "the insuperable line" Bentham wondered about is the capacity for moral reasoning. Our capacity for reason requires more of us. We appoint advocates to protect the interests of those unable either by act (e.g., criminals) or circumstance (e.g., children, insane, intellectually disabled). Perhaps our knowledge requires a similar advocacy for nature and the future.

## Who Owns the Environment?

The challenge of sustainability lies at least in part in our ability to resolve some of the underlying principles of our society's worldview. … In a culture where the individual is so highly valued and property has become a measure of distinction, our environment has been discounted and significantly compromised, at least in part, as a result. Capitalism is based on private ownership of the means and output of production. Private property is fundamental to the concept and cannot be separated from it. While this treatment is not inherently in disagreement with these principles, these are concepts that inform our worldview and are, for the most part, unquestioned but that have significant implications for our relationship with nature. Among the challenges we face is finding a way to reconcile the culture of individuality and the paradigm of property with an emerging recognition of a need for sustainability and intergenerational equity.

The idea of private property extends from the conclusion that if things found in nature are to be of use to a person there must be some means of appropriating them before they can be of use. No one would act to improve a thing without first being sure that his or her effort would serve him or her in some way or that he or she would have a right to the product of his or her effort. …

Economists refer to things found in nature as free goods; that is, they have no value in their natural state. It is only when the apple is picked and made available that it has value—likewise for oil or lumber or anything in nature. The oil company pays for equipment and labor, for access over the land, but when he removes the oil from the ground he pays nothing for it; it is free. The sum of the acts—not the intrinsic value of oil—of finding, drilling, pumping, refining and distributing the oil is represented in its price on the marketplace. Free goods are not to be confused with the concept of public goods that are goods or services whose benefits are not limited to one person; if one benefits, everyone benefits. As implied by the term, public goods cannot belong to one person. Examples might be air or national defense.

This devaluation of things in nature is compounded by the concept of discounting in modern finance. Discounting is simply the belief that future value is probably going to be worth less than current value so value should be taken as soon as possible for the best possible return. This is especially true when dealing with natural resources. For example, an acre of trees to be logged will continue to grow, adding wood over time, but rather than wait the trees are taken as soon as they have reached a size worth the effort of the harvest. To wait is to risk disease, fire, interference from other competitors, or introduction of new materials or technologies that change the value of the lumber or any other possible problem. Harvest the trees as soon as possible, and risk is mediated and the value can be invested in ways that grow faster than trees. Future earnings will also be paid in currency that is worth less than it is today. To offset the devaluation in currency value the trees have to grow even faster. In this scenario the trees as they stand have no value and the trees in the future represent less than no value because of inflation.

The implications of these underlying concepts of property, free goods, and discounting are clear. In the purely economic view there is no value in the environment. Value is described in terms of rights to use property or in terms of value added to things taken from nature. The economic system does not readily acknowledge the systemic value of the environment. For example, wetlands were seen as useless so in the past farmers and public agencies had programs to drain and fill them. These activities were undertaken to make the land "productive"—that is, available for use as farmland or for development. Such wetland literally sold for next to nothing in places and was given away by the government in others.

We better understand the systemic role of wetlands in the life cycle of living things and in the purification of water and air. As it turns out, wetlands are among the most productive natural landscapes. The approximate values of wetlands include removing pollutants, filtering debris from aquatic habitats, providing valuable habitat, perhaps providing buffers from destructive floods. So for every acre of wetland lost, people are willing to accept lower water quality, corresponding losses of habitat, the associated loss of biodiversity, increased flooding risks, and high costs for clean water. The effects of these costs are real but often difficult to assess in terms an individual might appreciate, so the value is seen as intangible, of no use in the marketplace. Still, the costs mount; the implications add up.

The difficulty we face is ultimately that environmental values are usually not expressed in economic terms. The two systems, economy and environment, are treated as if they exist apart from one another, distinct, unaffected by one another, but of course they are not. Sustainability in the end is a means of conducting environmentally sound economic activity for socially desirable outcomes. Too often environment and economy are considered as diametrically opposed to one another: a question of clean environment or healthy economy. There is nothing inherent in either, however, that precludes or excludes the other. The challenge lies in our definitions of quality. Economists are concerned because it is our economic behavior that must accommodate the change. Sustainability requires a long view. If we are to become sustainable the scope of our economic considerations must extend to include our impact on the environment and on the future. In short, we must begin to incorporate the externalities of our economic activity into the costs and to undertake some consideration of intergenerational equity. Externalities are those costs or benefits of a good or service not included within its price. A negative externality might be the costs of health care or lower property values imposed on a community by a polluting factory. Clearly, concepts such as free goods and discounting must be evaluated with some thought to the future as well as the short-term environmental impact. Accounting for the externalities of our actions is among the most basic principles of our sustainable future. ...

## The Value of Property and Nature

To consider the environmental impacts of the use and development of land without consideration of the economic implications is a pointless effort. In the end sustainability will be expressed as a balance of values. To date only the economic values have been considered. As we go forward it is necessary to capture the economic and environmental externalities so that this balance can be achieved. It is critical to understand both sides of the sustainability ledger if our efforts are to succeed.

Land development in the United States represents considerable economic value. Housing real estate value less mortgage expense was estimated to be about $18 trillion in 2006, and while there have been losses estimated to be around 20% in the years that followed, this still represents the largest pool of personal wealth in the United States.[32] Commercial real estate holdings are valued at about $5.3 trillion.[33] Public policy is predictably geared to protect and encourage this accumulation of wealth. Associated with the value of real estate is the role development plays in the gross domestic product (GDP). Construction alone contributes between 4% and 5% of the annual U.S. GDP. There is considerable economic investment and equity in real estate and development. Real estate, transfer, and development taxes all are major sources of local and state revenues. Our understanding of property valuation within a marketplace is fairly sophisticated, and we often hear issues and proposals equated to us in terms of negative or positive impacts on property values.

In light of this, how are we to find comparable value in nature? In 1973 economist Colin Clark demonstrated that by using standard economic models the best economic choice for whalers was to hunt the blue whale to extinction as quickly as possible. This choice would provide the whalers, and by extension the human race, the greatest possible return. The cost of limiting harvests until blue whale populations recovered and then hunting only at a sustainable level would yield less profit than killing them off and investing the profit. Clark thought that there was an unmistakable flaw in the model. Edward O. Wilson recounts Clark's argument: "The dollars-and-cents value of the blue whale was based only on the measures relevant to the existing market—that is, on the going price per unit weight of whale oil and meat. There are many other values, destined to grow along with our knowledge ... as science, medicine and aesthetics grow and strengthen, in dimensions and magnitudes still unforeseen."[34] Wilson and others raise the question as to whether contemporary economic models are able to evaluate or express the value of nature since they do not account for future value or current value beyond the utility of the products from a dead blue whale. The models all discount future value as previously described.

It is not that economists cannot put a value on natural resources. There are many models for the valuation of iron ore not yet extracted

or oil yet to be pumped. The reason for the inability of economic models to consider or account for the value of other resources is precisely that those resources are not traded. The marketplace, the willingness to pay, establishes value. The concept of willingness to pay is a measure of how desirable a thing is to possess, to own. What was the value of a blue whale in 1970 in the ocean? Zero. Nothing. What is not possessed has no value. What might the value of the same whale be in 100 years? 300 years? Difficult to say.

To identify the value of a thing in nature, unpossessed and unowned, other means of assessing value are necessary. Willingness to pay is a fundamental economic concept that is essentially a measure of the perceived utility of a good or service; in other words, one way to impute the value of an unpriced good or service is to determine how much someone is willing to pay for it.[35] If it was as simple as this, however, the model of discounting future value could not be refuted since only market utility is measured. Mark Sagoff suggests that the models needs to be reassessed: Willingness to pay "is not a value or a definition of value or a reason to value anything. ... To find out what we are willing to pay for, we have to determine what we value, not the other way around."[36]

We do find intrinsic value in nature that does not translate simply to market value. Value as resources yet to be identified, systemic values, beauty and solace come to mind. If these things are of value to us, willingness to pay may simply be inadequate measure and incapable of expressing the value of the resource.[37] We understand nature's value to transcend simple economic expression, but we are frequently faced with the proposition of tangible economic impacts weighed against transcendent values of nature. The economic values of nature are real but difficult to assess and to make real in the same way as real estate values.

## Reconciling Private Property and Sustainability

Capitalism is based on private property because private property is valued in the marketplace. Our current understanding is that the right to property ownership is a fundamental extension of our claim to individuality.[38] All things within this system, however, are valued according to their utility or to the willingness of a person to pay, but we recognize that all things valued cannot necessarily be expressed in those terms. To account for nature in a sustainable transaction will require different models and new considerations. The capacity to develop those models, however, will ultimately be a function of which theory of justice is chosen: utilitarian, libertarian, or Rawlsian.

Utilitarians will have to consider the greatest utility in terms of the environmental consequences of behavior and decisions and whether such long-term considerations can outweigh more short-term satisfaction. Can the satisfaction of the unborn be balanced against the living? Libertarians must consider whether the rights of future generations represent an encroachment on the rights of the current generation. Rawlsian justice must fold the interests of future into those considerations within the veil of ignorance. If the interests of the future and the intrinsic interests of the earth are considered, how will they be accommodated within these schools of thought?

There have been a number of efforts to describe nonmarket environmental values.[39] These methods range from calculating the systemic value if specific environmental functions (e.g., wetland capacity to clean water), future resource value (e.g., biotechnology value of unidentified species), and indirect value of natural resources (e.g., tourism value of forests or rivers). Interest in these proposals has been largely academic because there is insufficient public interest or awareness to compel the change. ...

Weak sustainability allows impact to be exchanged for equity, but that implies value that might not exist in the future. Strictly speaking weak sustainability may be difficult to support.[41] Though it appeals to economists it does not pass the tests of good science; it is inconsistent with what we know to be true about the natural world and the long-term interests of humanity. Economics as it is currently formulated is based on activities conducted in isolation from their impacts and extended costs.

Strong sustainability, however, limits our ability to act at all. Since we cannot know the future, we cannot risk any action that would have an impact. In this case the present is always a hostage of a future that doesn't even exist. Clearly, neither argument is entirely persuasive. Actions must be made after balanced considerations. The difficulty lies not in reconciling science and economics, however, but in understanding our values and putting them into practice. ...

## Endnotes

1. Berry, Thomas, *The Dream of the Earth* (San Francisco: Sierra Club Books, 1988), p. 52.

2. Beaucamp, Tom L. and Norman Bowie, *Ethical Theory and Business* (Upper Saddle River, NJ: Prentice-Hall, 1997), p. 34.

3. McMahan, Jeff, "Moral Intuition," in *The Blackwell Guide to Ethical Theory*, edited by Hugh LaFollette (Malden, MA: Blackwell Publishers, 2000), p. 93.

4. For the sake of time and space, I am going to limit my comments to the prevailing culture derived from Western Europe and the western Judeo-Christian experience. It should be noted, however, that many—in fact most—other cultures, existing and past, did incorporate nature. Nature in these cultures was inseparable from the culture, the day-to-day life, the very idea of being a person. This short chapter

limits itself to the dominant world culture and its recent awareness of the need to accommodate the laws of nature. Also, I will not provide the laundry list of environmental problems and challenges that serve to increase our awareness, though the list is compelling.

5. Kant, Emmanuel, *Foundation of Ethics*, translated by Leo Rauch, Agora Press, 1995, pp 47.

6. Rawls, John, *Justice as Fairness, A Restatement*, Belknap Press of Harvard University, Cambridge, 2001, pp. 14–18.

7. Gardner, Steve, Jeremy Bentham and John Stuart Mill, http://business.baylor.edu/Steve_Gardner/Bentham_Mill.htm

8. Barcalow, Emmett, *Moral Philosophy: Theories & Issues*, 2d ed. (New York: Wadsworth Publishing Co., 1998), pp. 202–209.

9. Locke, John, *Two Treatises of Government* (Cambridge, UK: Cambridge University Press, 2000), pp. 156–177.

10. Kant, Immanuel, *Lectures on Ethics* (Chicago: Louis Infield, 1980), pp. 239–240. Original lectures given between 1775 and 1780 were transcribed by three students and edited by Paul Menzer in 1924.

11. Des Jardins, Joseph, *Environmental Ethics, An Introduction to Environmental Philosophy*, 3d ed. (New York: Wadsworth Publishing Co., 2001), p. 221. Tertiary qualities such as beauty and inspiration were also discussed by early philosophers.

12. Baxter, William F., "People or Penguins," in *Moral Issues in Business*, 8th ed., edited by William H. Shaw and Vincent Barry (Belmont, CA: Wadsworth Publishing Co., 2001), pp. 580–584.

13. Ibid.

14. Ibid.

15. Ibid, p. 583.

16. Stone, Christopher D., Earth and Other Ethics: The Case for Moral Pluralism (New York: Harper &Row Publishers, 1988), pp. 73–83.

17. Singer, Peter, *Practical Ethics, Second Edition*, Cambridge, UK, Cambridge University Press, 1993, p. 52

18. Des Jardins, ibid, pg 114

19. Singer, Peter, *Practical Ethics, Second Edition*, Cambridge, UK, Cambridge University Press, 1993, p. 56

20. Singer, ibid., p. 59.

21. Singer, ibid., p. 60.

22. Singer, Peter, *Practical Ethics,* 2d ed. (Cambridge, UK: Cambridge University Press, 1993), p. 55.

23. Locke, John, *An Essay Concerning the True Original Extent and End of Civil Government*, 1690, Italics in original. 4LawSchool. Com online library, http://www.4lawschool.com/lib/locketable.htm

24. An Essay concerning the true original, extent and end of Civil Government (1690) from http://odur.let.rug.nk/~usa/D/1651-1700/locke/ECCG/govern02.htm, pg 26.

25. Ibid., p. 27.

26. Ibid., p. 31.

27. Ibid., p. 46.

28. Locke, John, *Second Treatise on Government,* Chp V, Sect 46., 1690, http://www.constitution.org/jl/2ndtreat.htm

29. Steinberg, Theodore, *Slide Mountain or the Folly of Owning Nature* (Berkeley: University of California Press, 1995), p. 11.

30. Brubaker, Elizabeth, "The Ecological Implications of Establishing Property Rights in Atlantic Fisheries," *Environment Probe*, April 1996.

31. Gillroy, John Martin, *Justice and Nature, Kantian Philosophy, Environment Policy & The Law* (Washington, DC: Georgetown University Press, 2000), p. 40.

32. Poole, William, "Real Estate in the United States," Remarks made to the Industrial Asset Management Council Convention, January 9, 2007, http://www.stlouisfed.org/news/speeches/2007/10_09_07.html

33. Standard and Poor's Commercial Real Estate Indices, December 31, 2007, http://www2.standardandpoors.com/spf/pdf/index/SP_GRA_Commercial_Real_Estate_Indices_Factsheet.pdf

34. Wilson, E. O., "What Is Nature Worth?" *Wilson Quarterly*, Winter 2002, p. 28.

35. Turner, R. Kerry, David Pearce, and Ian Bateman, *Environmental Economics, an Elementary Introduction* (Baltimore: Johns Hopkins Press, 1993), p. 108.

36. Campos, Daniel G., "Assessing the Value of Nature: A Transactional Approach," *Environmental Ethics*, 24(1), Spring 2002, p. 71.

37. Ibid.

38. The moral justification of capitalism is interesting, particularly with regard to our assumption that it is moral simply because it is so familiar to us. This is an interesting element but is beyond the scope of this inquiry.

39. Turner et al., *Environmental Economics*, p. 111.

40. Wilson, *Consilience* (New York: Alfred Knopf, 1998), p. 198.

41. Ayres, Robert U., Jeroen C. J. M. van den Bergh, and John Gowdy, "Strong versus Weak Sustainability: Economics, Natural Sciences and 'Consilience,'" *Environmental Ethics* 23(2), Summer 2001, p. 166.

42. Susskind, Charles, "An Engineer's Hippocratic Oath," in *Understanding Technology* (Baltimore: Johns Hopkins University Press, 1973), p. 118.

43. Obligation of an Engineer, University of Minnesota Duluth, College of Science and Engineering, http://www.d.umn.edu/cse/ooerc/oath.html

44. Albery, Nicholas, "Hippocratic Oath for Scientists," http://www.globalideas-bank.org/BOV/BV-381.HTML

45. Spier, Raymond, *Ethics, Tools and the Engineer* (New York: CRC Press, 2001).

46. Leopold, Aldo, "The Land Ethic," as quoted in "The Philosophical Foundations of Aldo Leopold's Land Ethic by Ernest Partridge," http://www.igc.org/gadfly/papers/;eopold.html

47. Ladd, John, "Collective and Individual Moral Responsibility in Engineering: Some Questions," in *Ethical Issues in Engineering*, edited by Deborah Johnson (Englewood Cliffs, NJ: Prentice-Hall, Inc., 1991), pp. 26–39.

# ENVIRONMENTAL LAWS AND SUSTAINABILITY

## An Introduction

By John C. Dernbach and Joel A. Mintz

### INTRODUCTION

Sustainable development provides a framework for humans to live and prosper in harmony with nature rather than living, as we have done for centuries, at nature's expense. Nonetheless, sustainability does not now have an adequate or supportive legal foundation, in spite of the many environmental and natural resources laws that exist. If we are to make significant progress toward a sustainable society, much less achieve sustainability, we will need to develop and implement laws and legal institutions that do not now exist, or that exist in a much different form. Since their clients in government, business, and nongovernmental organizations increasingly demand legal work that addresses sustainable development issues, lawyers have now begun to respond to that demand.

To achieve sustainability, we also need to recognize that, while environmental law is a key to achieving sustainability, it is only part of the necessary legal framework. Other needed legal involve a wide range of other laws, including land use and property laws, tax laws, laws involving our governmental structure, and the like. ...

### LESSONS ABOUT WHAT "LAW FOR SUSTAINABILITY" INVOLVES

The following is a suggested list of what "law for sustainability" involves. ... (Like many authors we use sustainability and sustainable development interchangeably.)

#### Use of Law to Require Integrated Decision-Making

The central action principle of sustainable development is integrated decision-making—the incorporation of environmental, social, and economic considerations and goals into

Note: this chapter has been excerpted from the original publication.

decisions. Nicholas Ashford and Ralph Hall argue that national governments in particular need to integrate "environmental, health, and safety regulation with industrial, trade, and employment policies" ([1], p. 282). Doing so will achieve not just incremental improvements but rather disruptive or breakthrough innovations that will, for example, improve the efficiency with which materials and energy are used by a factor of five or ten, and "foster significant opportunities for stable, rewarding, and meaningful employment with adequate purchasing power" ([1], p. 285). The integration of multiple national objectives necessarily opens up more space to solve problems. Integration fosters technological innovation because a greater range of mutually reinforcing objectives can present a larger number of options for achieving these objectives. Governments can achieve this kind of policy integration, Ashford and Hall argue, through a variety of legal and policy tools, including the use of regulation to foster innovation; research and development; "removing regulatory barriers to innovation;" tax policies; and encouragement of management-labor bargaining "before technological changes are planned and implemented" ([1], p. 287).

Integration across levels of government is also important. In many cases, the legal rules for a particular sustainability objective at a higher level of government are completely separate from the relevant legal rules at a lower level of government. As a result, the lower level of government can make decisions that frustrate or impede sustainability goals. Rachel Medina and A. Dan Tarlock explain how the state of California addressed just such a problem [2]. The state's 2006 climate change legislation of reducing California's greenhouse gas emissions

to 1990 levels by 2020, but it did not address local land use laws that were contributing to sprawl and greater greenhouse gas emissions. In 2008, after litigation that challenged these local laws, the legislature passed a law encouraging plans that provide for less automobile use.

## Use of Pre-Existing Laws to Foster Sustainability

Most observers recognize that sustainability requires new laws and modifications to existing laws. It is less often recognized that sustainability can be achieved by simply applying existing laws to new problems, or by making incremental changes in those laws.

Many but not all of these laws are traditional environmental laws. Nongovernmental organizations and the California Attorney General, for example, forced municipalities in that state to consider the greenhouse gas emission impacts of their land use decisions by filing and settling lawsuits under the 1970 California Environmental Quality Act (CEQA) [2]. This statute requires state and local governments to prepare environmental impact reports for decisions that will have a significant impact, and to reduce or avoid impacts whenever feasible. This litigation, as already noted, led to the adoption of a 2008 statute describing local responsibilities to limit greenhouse gas emissions with greater particularity than CEQA. But that legislation would not likely have been adopted without the CEQA litigation.

Similarly, Robin Craig and J. B. Ruhl survey a range of legal and policy tools that can be used to foster sustainable management of coastal ecosystems—integrated, place-based management strategies and innovative regulations, including market-based instruments. Their starting point is existing law, and they show how these existing laws could be modified to achieve more sustainable outcomes, including adaptation to climate change. For example, they advocate wider use of collaborative governance structures, greater use of "reflexive law" (such as information reporting), and more use of economic incentives. While they recognize that governments have begun to use these tools, they argue that they could be "used more pervasively and creatively as part of new sustainable governance institutions" ([3], p. 1380).

Many of the laws that provide starting points for sustainability are not environmental laws. A 1994 Cuban law made it easier for residents of urban areas to grow and sell food on unused land. Since then, the government has provided financial, technical, and marketing assistance for this kind of agriculture. The overall program has increased access to food, led to more organic and sustainable food production, and created jobs—all key elements of a food security program—even though much food is sold on the black market or through foreign currency. Cuba has one of the world's most aggressive and sustainable urban agriculture systems, and it is developing in stages that began with the 1994 law ([4], pp. 1703–1705).

Property law, particularly the law involving common-interest communities (known as homeowner associations in a residential context) can also be used to foster sustainability. Anthony Schutz proposes that property law be applied in the context of the ranchland in the U.S. Great Plains to take advantage of consumer demand for horseback riding, camping, wildlife observation, hiking, and hunting. In this region, only 2% of the land is managed for biodiversity conservation while three fourths of the land is in private ownership ([5], p. 2321). Schutz's

article identifies many of the ways that groups of private landowners could use the law of common-interest communities to organize themselves to provide these services collectively. While the organization of such communities would require landowners to make a great many decisions—including what they want to produce as a group, and how they would allocate the benefits—these communities would provide economic opportunities that would benefit the environment as well.

Nor do these existing laws need to be from the same jurisdiction; comparative law has a significant role to play in the quest for sustainability. Kenneth Abbott and Gary Marchant review five possible mechanisms for institutionalizing sustainability across the entire federal government in the United States. Each of these mechanisms is based on an existing law from the U.S. or another jurisdiction. For each, they identify ways in which an existing law could be modified, strengthened, or extended to foster sustainability at the national level. As they note, the incremental adaptation of existing laws means that the innovations they represent are achievable ([6], p. 1938). Similarly, in analyzing an appropriate legal structure for sustainability in the United Kingdom, Andrea Ross draws on examples from Wales, Scotland, and the Canadian provinces of Manitoba and Quebec [7].

Existing laws might also be effectively supplemented with other approaches, including community—based social marketing techniques. As Amanda Kennedy explains, "[c]ommunity-based social marketing aims to produce behavioral change via direct communication and community level initiatives, concentrating upon removing barriers to change" ([8], p. 1139). These marketing techniques include public commitments, incentives, public communication strategies, prompts to encourage actions that people might not otherwise do, and social norms. A considerable body of social science research demonstrates that community-based social marketing can cause individual behavior to be more sustainable. She provides examples where social marketing appeals made to individuals have resulted in more effective environmental regulation—including air pollution reduction in Portland, Oregon, and water use restrictions in Kamloops, British Columbia. Environmental regulation and community—based social marketing together, Kennedy concludes, are both more effective and more cost effective than either one by itself.

## The Centrality of Sub-National Governments in Achieving Sustainability

Across the world, sub-national units of government—such as provinces, states and local governmental units—have substantial, if not primary, legal responsibility for issues central to sustainable development. These include, but are not limited to, education, provision of drinking water and sewage disposal, and land use control. This is the case not only in nations with a federal form of government, but also in countries like the United Kingdom, where there has been substantial devolution of decision-making authority (including authority for sustainable development strategies) to sub-national governments (e.g., Scotland, Wales, Northern Ireland) ([7], pp. 1106–1107). In view of this, much of the progress necessary to achieve sustainability, as well as the most important obstacles to its achievement, are located at the sub-national level. However, when responsibility

is shared to some degree between national and sub-national levels, a lack of national action on important issues (e.g., climate change) may prompt sub-national governments to address that issue on their own [2].

Apart from legal responsibility, the more granular scale of decision making at more local levels corresponds to the more granular scale of the problem. Better local land use and sustainability planning and decision-making can make communities significantly more livable because that is the level at which many of the relevant challenges occur, and also the level at which decisions are made ([2], pp. 1756–1757). For example, Maja Goepel points out that Tuscany's genetic heritage law allows the listing and limited exchange of local breeds and varieties of crop plants, thus identifying them and providing an incentive to protect them ([4], pp. 1702–1703). Such a law would likely be much harder to implement at the national level in Italy.

At the same time there are limits to the efficacy of decision-making at sub-national levels of government. Thus, Robert Adler suggests that greater frequency and severity of droughts because of climate change may force national governments to assume more responsibility for issues like water sustainability. Where that occurs, sub-national governmental units (such as U.S. states), where most water law decisions are made, are likely to end up with fewer responsibilities [9].

## The Background Law of Unsustainable Development

An enormous but not fully characterized problem is the extent to which the law now works *against* sustainability. Environmental laws are easy to identify because they are usually labeled as such. By contrast, there are few if any explicitly anti-environmental or anti-sustainability laws. Instead, such laws deserve to be included in the law of unsustainable development because of their effects. The articles in this special issue provide ample evidence of the existence of such laws. Land use laws are part of the problem, as Medina and Tarlock explain. "United States land-use law encourages cities not to constrain residential and commercial land-use choices but to expand the range of these choices by favoring low density, suburban urbanization characterized by economic, racial and social segregation" ([2], p. 1745). Similarly, Robert Adler argues, "existing legal regimes address drought in a largely reactive way, fail to promote sustainable systems and practices that might help to mitigate or prevent drought impacts, and in some ways actually decrease sustainability and hence Increase society's vulnerability to drought" ([9], p. 2177). Ashford and Hall argue, more strongly, that the unsustainable beneficiaries of the present system are supported and encouraged by a variety of laws, and that they use these laws and existing legal institutions to resist change ([1], p. 289).

## The Growing Importance of Climate Change

The importance of climate change to sustainability has grown rapidly in recent years. … They focus on the impact of local decisions on climate change, and the need to reduce greenhouse gas emissions that would otherwise result from those decisions ([2], pp. 1743–1747). Of equal importance, but much less understood, is a need to modify pre-existing laws to adapt to climate change. Adler believes that laws relating to drought—a recurring issue in human

history—will likely need to be reexamined in light of the more frequent and severe droughts that are projected to occur because of human-induced climate change. "An increased focus on sustainability could improve the ability of affected communities to anticipate, cope with, and even prevent drought impacts, he says ([9], p. 2180). For Adler, this increased focus on sustainability would include a new legal approach to drought based on several factors, including a social decision "about how the risk of water shortages should be distributed," legal promotion of the sustainability of water supplies and allocation of water to places where it is most needed, integration of drought planning into water resources law and policy (rather than treating drought as a separate issue), and economic and other incentives for sustainable water use. However, changes to water law by themselves are not enough. A truly sustainable approach to water policy will also need to "re-evaluate and amend laws and policies that drive important underlying economic decisions by the agriculture industry and others" ([9], p. 2192).

## The Need to Use Law to Protect and Restore Ecological Integrity

While sustainable development is commonly understood as involving the relationship between the social, environmental, and economic domains of human existence, Klaus Bosselmann insists on the primacy of protecting and restoring ecological sustainability [10]. There needs to be an ecological bottom line, he argues, because "the ecological basis for human survival is at risk" ([10], p. 2442). Environmental law has not prevented the continuation of widespread environmental degradation around the world; the "global commons—climate, biodiversity, the oceans—are in rapid decline and the human ecological footprint both in absolute terms and *per capita* is getting larger" ([10], p. 2433). Bosselmann suggests amending New Zealand's well-known Resource Management Act so that its purpose would be "to achieve ecological sustainability in New Zealand." While changes such as this would not automatically "set us on a sustainable path," he argues, they could at least "provide the direction for such a path" ([10], p. 2442).

The protection and restoration of ecosystems is also a subtext. ... For example, Schutz's proposal to use the law of common-interest communities on lands that are now devoted primarily to livestock grazing is specifically intended to increase biodiversity conservation on those lands [5].

## The Importance of Judicial Review and Nongovernmental Organizations

Sustainable development requires public participation and effective access to the courts. Without an engaged public and an independent judiciary, laws on the books may not be implemented or enforced. ... The California litigation that ultimately forced local governments to consider their climate change impacts was originally brought by a nongovernmental organization, the Center for Biological Diversity ([2], p. 1749). Although the case was settled rather than decided by a court ([2], pp. 1754–1755), the availability of an effective court system provided impetus for the settlement.

Goepel uses her article to advance the work of the World Future Council, the nongovernmental organization for which she works. Her article, which itself builds on the work of the International Law Association, another nongovernmental organization, illustrates the importance of such organizations in developing,

refining, and advocating legal and policy tools for sustainability [4].

## The Need to Translate Sustainability into Specific Legal Principles

While the vagueness of sustainable development is widely lamented, the international agreements on which sustainability is based as well as a growing body of scholarship and experience have set out basic legal principles for sustainability. Craig and Ruhl identify eight such principles: the polluter-pays principle, the use of best available science, the precautionary principle, intergenerational sustainability, transnational sustainability, accounting for ecosystem services, integrated decision-making, and adaptive management. They explain how those principles can be applied to the sustainable management of coastal ecosystems ([3], pp. 1367–1375).

Somewhat similarly, Bosselmann insists on the necessity of ecological sustainability as a bedrock sustainability principle for law. He would define ecological sustainability as "preservation or restoration of the integrity of any ecosystem in the biosphere," and would define ecological integrity as "the ability of an ecosystem to recover from disturbance and re-establish its stability, diversity and resilience" ([10], p. 2441).

There also have been formal efforts to state principles of national governance that could be applied to sustainability. Goepel describes a project by the International Law Association to set out seven specific principles (sustainable use of natural resources; equity and the eradication of poverty; precautionary approach; public participation and access to justice; good governance; common but differentiated responsibilities among developed and developing nations; and integration of human rights and social, economic, and environmental objectives) ([4], pp. 1696–1698). Goepel observes that these principles can also be applied to national governance. Ross explains that the United Kingdom's Sustainable Development Commission has used three broad criteria that are essential to make sustainable development happen at the national level. These are public understanding of the big picture; a comprehensive framework for integrating conflicting priorities; and a toolkit of policies, practices, and laws for implementing sustainability ([7], pp. 1104–1105).

## The Challenge of Creating an Appropriate National Legal Structure for Sustainability

The sustainable development commitments that nations made at the United Nations Conference on the Environment and Development in Rio de Janeiro in 1992, and that they reaffirmed in 1997 and 2002, are, of course, *national* commitments. The cross-cutting nature of sustainability means that it is not confined to a single subject or a single administrative agency, department, or ministry. Rather, national commitments require the engagement of the federal or national government as a whole. …

The United Kingdom has had a national sustainable development strategy since 1994, the content of which has evolved over time. Nonetheless, that strategy lacks a strong legal foundation. The United Kingdom's strategy is progressive in tone and substance, according to Ross. However it has not been particularly effective at delivering the three criteria (mentioned above) that the Sustainable Development Commission says must be met. The biggest problem, as Ross sees it, is that "there seems to be very little understanding or coherent thought about what exactly sustainable development means and its role in governance" ([7], p. 1109). Three models for legislation are available—a procedural model

(requiring, for example, the development of a strategy but not necessarily requiring adherence to the strategy), a law that explicitly establishes a sustainable development strategy as the point of reference for all decision-making (or, at a minimum, environmental decision-making), and a law that makes sustainable development the organizing principle for national governance. Although she prefers the third model, Ross advocates that the adoption of legislation be staged over time, beginning with the procedural model, because the implementation of sustainable development in the United Kingdom since 1994 has also evolved in stages. Whichever legal approach is used, she concludes, will cause sustainability to be taken more seriously, create legally enforceable obligations (if only to observe required procedures), and attract greater public attention to sustainability ([7], p. 1121).

Abbott and Marchant argue that at least five mechanisms are available in the United States to institutionalize a national approach to sustainability ([6], pp. 1926–1936). These are: (1) an executive order requiring federal agencies to work broadly toward sustainability, (2) a sustainability impact assessment process that would include analysis of the effect of agency policies and programs, (3) a nonpartisan Congressional Joint Committee on Sustainability to assist Congress in examining the effect of laws and recommending reforms, (4) a federal Sustainability Commission modeled on the United Kingdom Sustainable Development Commission that would advise the government and advocate and monitor sustainability activities, and (5) an independent Sustainability Law Reform Commission that would be "tasked with reviewing existing federal law from the perspective of sustainability and recommending amendments, enactments and repeals" ([6], p. 1932). These mechanisms, they say, illustrate a reasonable range of possible approaches that could be used in the United States to foster sustainability.

## The Importance of Sustainability Assessment Tools and Institutions Before and After Laws Are Adopted

It is not enough, of course, to simply throw laws at a problem. Laws must be designed and drafted with care to achieve particular results, and they must be evaluated carefully afterwards to see if they have actually achieved the desired results. For sustainability laws, the availability of credible and widely applicable assessment tools and institutions is especially important.

These tools can include existing social science research methodologies. For example, Kennedy notes many social science studies evaluating the effectiveness of environmental regulations that are combined with community-based social marketing [8]. Nevertheless, new tools are also required. Using food security laws as an example, Goepel describes how the International Law Association's seven principles can be used to assess the sustainability of particular legal measures [4]. The assessment methodology that Goepel favors translates these seven principles into interview questions to be posed to government officials, members of nongovernmental organizations, and academic and other experts. The methodology that she recommends is empirical, not theoretical, and qualitative, rather than quantitative. After using this method to evaluate food security laws in Tuscany, Italy; urban Cuba; and Belo Horizonte, Brazil, Goepel concludes that "the seven principles, even though originally defined for international law, have the potential to serve as an ideal norm that is universal and yet flexible enough to guide the drafting and amendment of laws and policies on any governance level" ([4], p. 1696).

Assessment tools are not enough, however. Without some institutional mechanism to assess laws from a sustainability perspective before and after they are adopted, sustainability assessment is likely to be sporadic and ineffectual. The purpose of many of the mechanisms identified by Abbott and Marchant is to assess laws from a sustainability perspective on an ongoing basis—both before and after these laws are adopted. At the same time however, those authors recognize that, standing alone, institutions for assessment would be insufficient without more precisely defined principles that can form the basis for assessment [6].

## The Importance of "Soft" Law

At the international level, scholars often distinguish between hard law (e.g., treaties that are in force) and soft law (e.g., declarations at international conferences). In practice, though, there is a sliding scale of "hardness" and "softness" at both the national and international levels; some hard laws contain soft requirements (e.g., to "consider" or "assess") while some soft law instruments contain hard law norms (e.g., legal principles). Abbott and Marchant analyze five mechanisms based on their relative hardness or softness, and conclude that "institutions with widely varying forms and levels of legal authority can make valuable contributions to sustainability law and policy; consideration should not be limited to institutions with—'hard' legal mandates" ([6], p. 1937). In fact, as Kennedy suggests, some mechanisms may not even qualify as soft law. Social norms, public communications, and other aspects of community-based social marketing nonetheless can also shift individual behavior in a more sustainable direction [8].

## Conclusion

The underlying challenges that sustainability addresses—widespread poverty and growing global environmental degradation—are urgent challenges. Law can make a variety of key contributions in achieving sustainability. We need to find ways to accelerate the use—and analysis—of laws that can foster sustainability. ...

## References

1. Ashford, N.A.; Hall, R.P. The importance of regulation-induced innovation for sustainable development. *Sustainability 2011, 3,* 270–292.

2. Medina, R.; Tarlock, A.D. Addressing climate change at the state and local level: using land use controls to reduce automobile emissions. *Sustainability 2010, 2,* 1742–1764.

3. Craig, R.K.; Ruhl, J. Governing for sustainable coasts: complexity, climate change, and coastal ecosystem protection. *Sustainability 2010, 2,* 1361–1388.

4. Goepel, M. Formulating future just policies: applying the Delhi sustainable development law principles. *Sustainability 2010, 2,* 1694–1718.

5. Schutz, A.B. Grassland governance and common-interest communities. *Sustainability 2010, 2,* 2320–2348.

6. Abbott, K.W.; Marchant, G.E. Institutionalizing sustainability across the federal government. *Sustainability 2010, 2,* 1924–1942.

7. Ross, A. It's time to get serious—Why legislation is needed to make sustainable development a reality in the UK. *Sustainability 2010, 2,* 1101–1127.

8. Kennedy, A.L. Using community-based social marketing techniques to enhance environmental regulation. *Sustainability 2010, 2,* 1138–1160.

9. Adler, R.W. Drought, sustainability, and the law. *Sustainability 2010, 2,* 2176–2196.

10. Bosselmann, K. Losing the forest for the trees: environmental reductionism m the law. *Sustainability 2010, 2,* 2424–2448.

Section V

# WATER, ENERGY, FOOD AND HEALTH

# WATER

## Adapting to a New Normal

By Sandra Postel

Water, like energy, is essential to virtually every human endeavor. It is needed to grow food and fiber, to make clothes and computers, and, of course, to drink. The growing number of water shortages around the world and the possibility of these shortages leading to economic disruption, food crises, social tensions, and even war suggest that the challenges posed by water in the coming decades will rival those posed by declining oil supplies.

In fact, our water problem turns out to be much more worrisome than our energy situation, for three main reasons. First, unlike oil and coal, water is much more than a commodity: It is the basis of life. Deprive any plant or animal of water, and it dies. Our decisions about water—how to use, allocate, and manage it—are deeply ethical ones; they determine the survival of most of the planet's species, including our own. Second, also unlike oil and coal, water has no substitutes. The global economy is transitioning away from fossil fuels toward solar, wind, and other noncarbon energy sources, but there is no transitioning away from water. And third, it is through water that we will experience the impacts of climate change most directly.

The rise in global temperatures driven by the last 150 years of humanity's greenhouse gas emissions is fundamentally altering the cycling of water between the sea, the atmosphere, and the land. Climate scientists warn of more extreme floods and droughts and of changing precipitation patterns that will make many dry areas drier and wet areas wetter. They warn of melting glaciers and ice caps that within a few decades could severely diminish the river flows upon which nearly a third of the world's people depend. As if on cue, nature seems to be highlighting these warnings at every turn:

- Floods, droughts, storms, and other climate-related natural disasters forced 20 million people from their homes in 2008.

107

- Australia remains locked in a decade-long drought deemed the worst in the country's 117 years of record-keeping.
- In late August 2008, India faced the dislocation of some 3 million people when the Kosi River breached a dam and roared out of the Himalayas, causing the worst flooding of that river in fifty years.
- Ten months later, India witnessed its driest June in eighty years with millions of farmers unable to plant their crops.
- In 2009, famine stalked millions in the Horn of Africa, as failed rains led to the worst food crisis in Ethiopia and Kenya in a quarter century.

The United States is by no means immune to these climate-related water risks. While farmers in the Midwest continued recovering from the spring flood of 2008 (in some areas, the second "100-year flood" in fifteen years), farmers in California and Texas fallowed cropland and sent cattle prematurely to slaughter to cope with the drought of 2009. In the Southeast, after twenty months of dryness, Georgia governor Sonny Perdue stood outside the state capitol in November 2007 and led a prayer for rain, beseeching the heavens to turn on a spigot for his parched state. Two years later, Perdue was pleading instead for federal aid after intense rainstorms caused massive flooding near Atlanta that claimed at least seven lives.

Although none of these disasters can be pinned directly on global warming, they are the kinds of events climate scientists warn will occur more often as the planet heats up. Even more worrisome, the effects of climate change are already calling into question the very assumptions that have underpinned water planning and management for decades. In 2008, seven top water scientists argued persuasively in the journal *Science* that "stationarity"—the foundational concept that natural systems vary and fluctuate within an unchanging set of boundaries—is no longer valid for our understanding of the global water system. In other words, when it comes to water, the past is no longer a reliable guide to the future. The data and statistical tools used to plan $500 billion worth of annual global investments in dams, flood-control structures, diversion projects, and other big pieces of water infrastructure are no longer trustworthy.

This is not just a problem for the planners and civil servants who run our local water systems. It raises very serious questions about community health, public safety, food security, and risk management. Will those levees keep the river within its banks? Should that expensive new dam be built when its useful life will be shortened by silt washed down from flooding mountainsides? Will farms get needed irrigation water once the glacier-fed river flows have dwindled? How do we guard against what once seemed unthinkable—the drying up of prime water sources?

In more and more regions of the world, the unthinkable seems to be close at hand. Many Australian water managers now believe that a decade-long dry spell that has sent rice production plummeting, depleted reservoirs, and left the Murray River trickling into the sand is not going away. Increasingly, the question "Down Under" is not when the drought will end, but how this country of more than 21 million people—and its globally significant agricultural sector—can adapt to a permanently drier climate.

In the U.S. Southwest, a similar day of reckoning is on the horizon. Scientists at the Scripps Institution of Oceanography at the University of California–San Diego estimate that there is a 50 percent chance that Lake Mead—the vast reservoir that delivers Colorado River water to tens of millions of people and one million acres of irrigated land—will dry up by 2021. In 2000, Lake Mead stood at 96 percent of capacity; by the summer of 2009, it was down to 43 percent. After analyzing nineteen climate models, a team of thirteen earth scientists concluded in 2007 that the Dust Bowl-like dryness seen in the region in recent years "will become the new climatology of the American Southwest within a time frame of years to decades." As in Australia, it may be folly—and a loss of precious time—to assume that business as usual and life as it's currently lived in the Southwest can continue.

The water challenges confronting us locally, regionally, and globally are unprecedented. They call for fundamental changes in how we use, manage, and even think about water. The good news is that it's within our economic and technological ability to have a future in which all food and water needs are met, healthy ecosystems are sustained, and communities remain secure and resilient in the face of changing circumstances. The path most of the world is on, however, will not lead to this more desirable state.

## WHERE WE ARE, AND HOW WE GOT HERE

At first glance, it's hard to believe the world could be in trouble with water. Ever since the Apollo astronauts photographed Earth from space, we've had the image of our home as a strikingly blue planet, a place of great water wealth. But from a practical standpoint, this image is largely illusory. Most of Earth's water is ocean, which provides a multitude of benefits but is far too salty to drink, irrigate crops, or manufacture computer chips. Only a tiny share of all the water on Earth—less than one-hundredth of 1 percent—is fresh and renewed each year by the solar-powered hydrologic cycle.

Although renewable, freshwater is finite: The quantity available today is virtually the same as when civilizations first arose thousands of years ago. As world population grows, the volume of water available per person decreases; thus, between 1950 and 2009, as world population climbed from 2.5 billion to 6.8 billion, the global renewable water supply per person declined by 63 percent. If, as projected, world population climbs to 8 billion by 2025, the water supply per person will drop by an additional 15 percent.

Though telling, these global figures mask the real story. The rain and snow falling on the land is not evenly distributed across the continents or throughout the year (see figure 7.1). Many of the world's people and farms are not located where the usable water is. China, for instance, has 19.5 percent of the world's population, but only 7 percent of the renewable freshwater. The United States, by contrast, has 4.5 percent of the world's population and nearly 8 percent of the renewable freshwater. Even so, most of U.S. farm irrigation and urban growth is in the West, which has much less water than the eastern United States.

For most of modern history, water management has focused on bringing water under human control and transferring it to expanding cities, industries, and farms. Since 1950, the number of large dams has climbed from 5,000 to more than 45,000—an average construction rate of two large dams per day for half a century. Globally, 364 large water-transfer schemes move 400 billion cubic meters (1 cubic meter equals about 264 gallons) of water annually from one river basin to another—equivalent to transferring the annual flow of twenty-two Colorado Rivers. Millions of wells tap underground aquifers, using diesel or electric pumps to lift vast quantities of groundwater to the surface. It's hard to fathom today's world of 6.8 billion people and $60 trillion in annual economic output without such water engineering. It has

FIGURE 11-1. SURFACE WATER STRESS AREAS OF THE WORLD.
The Nature Conservancy and World Wildlife Fund, Inc. http://www.feow.org/maps/threat/surface_water_abstraction_stress_to_rivers

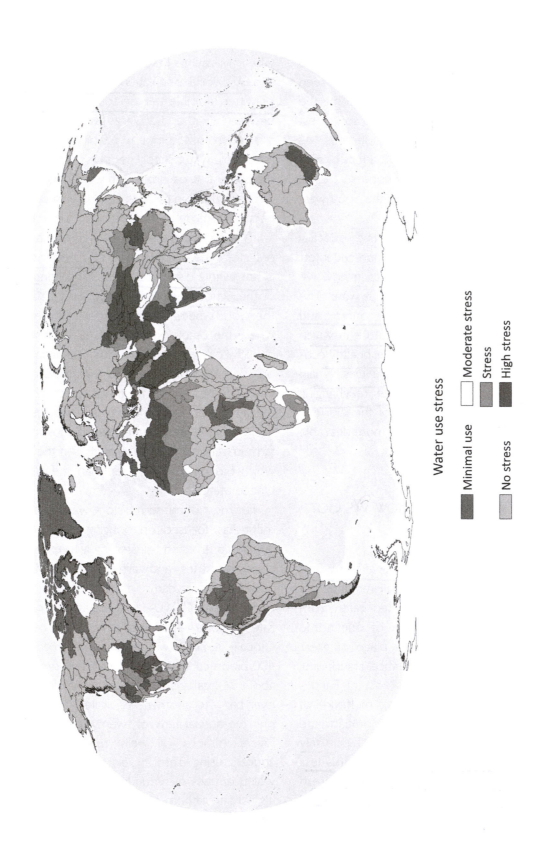

Water use stress

Minimal use     Moderate stress

No stress     Stress

    High stress

allowed oasis cities like Phoenix and Las Vegas to thrive in the desert, world food production to expand along with population, and living standards for hundreds of millions to rise.

But the benefits of water development have not been shared equitably. More than 1 billion people lack access to safe drinking water, and some 850 million people are chronically hungry. Moreover, many regions have overshot their sustainable limits of water use. An unsettling number of large rivers—including the Colorado, Rio Grande, Yellow, Indus, Ganges, Amu Darya, Murray, and Nile—are now so overtapped that they discharge little or no water to the sea for months at a time. The overpumping of groundwater is causing water tables to fall across large areas of northern China, India, Pakistan, Iran, the Middle East, Mexico and the western United States. As much as 10 percent of the world's food is produced by overpumping groundwater. This creates a bubble in the food economy far more serious than the recent housing, credit, or dot-com bubbles, for we are meeting some of today's food needs with tomorrow's water.

This overpumping is particularly serious in India. Using satellite data, scientists have recently estimated that groundwater is being depleted across northern India, which includes the nation's breadbasket, to the tune of 54 billion cubic meters per year. As wells run dry, the nation's food supply—as well as the livelihoods of the region's 14 million people—is increasingly at risk. Likewise, in the United States, the massive Ogallala Aquifer is steadily being depleted. The Ogallala spans parts of eight states, from southern South Dakota to northwest Texas, and provides 30 percent of the groundwater used for irrigation in the country. As of 2005, a volume equivalent to two-thirds of the water in Lake Erie had been depleted from the Ogallala. As in India, most farmers will stop irrigating when the wells run dry or the water drops so far down that it's too expensive to pump.

It is tempting to respond to these predicaments with bigger versions of the familiar solutions of the past—drill deeper wells, build bigger dams, move more river water from one place to another. Indeed, many leaders and localities are responding in just that way. By some estimates, the volume of water moved through river-transfer schemes could more than double by 2020.

China is proceeding with a massive $60 billion project to transfer water from the Yangtze River basin in the south to the water-scarce north. If completed, it would be the largest construction project on Earth, transferring 41.3 billion cubic meters of water per year—a volume equal to half the Nile River. India's Interlinking Rivers Project would be even more grandiose. Estimated to cost at least $120 billion, it entails building 260 transfers between rivers with much of the water moved from northern Himalayan rivers, including the Ganges and Brahmaputra, to water-scarce western provinces. Though still in the planning stages, the main goal would be to expand the nation's irrigated area by about a third, some 35 million hectares.

In a world of changing rainfall patterns and river flows, substantial hydrologic uncertainty, and rising energy costs, such mega-projects are risky. They often take decades to complete, so payback periods on the large capital investments can be very long (if full payback occurs at all). They often worsen social inequities, such as when poor people are dislocated from their homes to make way for the dams and canals and "downstream" communities lose the flows that sustained their livelihoods. And serious environmental damage—from soil salinization, water waste, altered river flows, and the loss of fisheries—routinely follows on the heels of such projects. Moreover, large-scale infrastructure built to accommodate river flows today may be poorly matched to climate-altered flows of the future. The Himalayan rivers central to India's

Interlinking Rivers Project, for example, will carry greatly diminished flows once the glaciers that feed them disappear.

In addition, giant water projects require giant quantities of energy. Pumping, moving, treating, and distributing water take energy at every stage. Transferring Colorado River water into southern California, for example, requires about 1.6 kilowatt-hours (kWh) of electricity per cubic meter of water; the same quantity of water sent hundreds of kilometers from north to south through California's State Water Project takes about 2.4 kWh. As a result, the energy required to provide drinking water to a typical southern California home can rank third behind that required to run the air conditioner and refrigerator.

Another increasingly popular option for expanding water supplies—desalination— imposes a high energy price as well. Producing 1 cubic meter (about 264 gallons) of drinkable water from saltwater through reverse osmosis requires about 2 kWh of electricity, usually produced from fossil fuels. Although that energy requirement is down from 5–10 kWh twenty years ago, it is still energy intensive. Moreover, today's most energy-efficient desalting plants are approaching the theoretical thermodynamic limit for separating salts from water, so further energy reductions will be modest at best. Currently, the roughly 15,000 desalination plants worldwide have the capacity to produce 15.3 billion cubic meters of water per year, which is less than 0.5 percent of global water demand. Some 47 percent of this capacity is in the Middle East, where many nations can afford desalination—essentially turning their oil into water.

Despite desalination's high costs, carbon dioxide output, risks to coastal marine environments, and production of toxic waste, global capacity roughly doubled between 1995 and 2006. Most U.S. capacity is in Florida, California, and Texas, with many more plants slated to be built. Unfortunately, planners and policy-makers still eyeing desalination as a silver-bullet solution to water shortages apparently miss—or dismiss—the perverse irony: By burning more fossil fuels, desalination will likely worsen the problem they are trying to solve while making local water supplies more and more dependent on increasingly expensive fossil fuels.

## A SMARTER PATH TOWARD WATER SECURITY

As with many challenges, finding the best solutions requires first asking the right questions. Typically, when planners and engineers see a water shortage on the horizon, they ask themselves what options exist to expand the supply. The typical answer: Get more water from a distant river, deeper wells, or a desalination plant.

But as the limitations of these "supply-side" options have become more apparent, a vanguard of citizens, communities, farmers, and corporations has started asking a different question: What do we really need the water *for*, and can we meet that need with less? The upshot of this shift in thinking is a new movement in water management that is much more about ideas, ingenuity, and ecological intelligence than it is about pumps, pipelines, dams, and canals.

This smarter path takes many forms, but it embodies two strategic attributes. First, solutions tend to work with nature, rather than against it. In this way, they make effective use  of so-called ecosystem services—the benefits provided by healthy watersheds, rivers, wetlands, and other ecological systems. And second, through better technologies and more informed choices, these solutions seek to raise water productivity—the benefit derived from each liter of water extracted from a river, lake, or aquifer.

Working with nature is critically important to building resilience and reducing the energy costs associated with water delivery and use. We can think of a landscape composed of well-functioning ecosystems as "green infrastructure" that provides valuable services to society, just as roads and bridges do. Healthy rivers and watersheds, for instance, filter out pollutants, mitigate floods and droughts, recharge groundwater supplies, and sustain fisheries. They do this work with free energy from the sun—no fossil fuels or manufactured energy is required. By contrast, all the technological alternatives—building and running a treatment plant to remove pollutants, artificially recharging groundwater, constructing dikes and levees, raising fish on farms—require external inputs of increasingly expensive energy.

Of course, one of the most important "services" healthy watersheds perform is the provision of clean drinking water. If a watershed is doing the work of a water treatment plant—filtering out pollutants, and at a lower cost to boot—then it often pays to protect that watershed. New York City, for instance, is investing some $1.5 billion to restore and protect the Catskills-Delaware watershed (which supplies 90 percent of its drinking water) in lieu of constructing a $6 billion filtration plant that would cost an additional $300 million a year to operate. A number of other cities across the United States—from tiny Auburn, Maine, to the city of Seattle—have saved hundreds of millions of dollars in avoided capital and operating costs by opting for watershed protection over filtration plants. In doing so, they have enjoyed many other benefits, such as preserving open space, creating recreational opportunities, protecting habitat for birds and wildlife, and (by preserving trees) mitigating climate change.

Other innovative ideas are coming from Latin America, where some cities are establishing watershed trust funds. For instance, Rio de Janeiro in Brazil collects fees from water users to pay upstream farmers and ranchers $71 per hectare ($28 per acre) to protect and restore riparian forests, safeguarding the water supply and preserving habitat for rare birds and primates. A public watershed protection fund in Quito, Ecuador, started in 2000 in partnership with the Nature Conservancy, receives nearly $1 million a year from municipal water utilities and electric companies. Quito's water fund has become a model for other Latin American cities, including Cuenca, Ecuador, and Lima, Peru.

There are many ways communities can work with nature to meet their water needs while reducing energy costs and building resilience. Communities facing increased flood damage, for instance, might achieve cost-effective flood protection by restoring a local river's natural floodplain. After enduring nineteen flood episodes between 1961 and 1997, Napa, California, opted for this approach over the conventional route of channelizing and building levees. In partnership with the Army Corps of Engineers, the $366 million project is reconnecting the Napa River with its historic floodplain, moving homes and businesses out of harm's way, revitalizing wetlands and marshlands, and constructing levees and bypass channels in strategic locations. In addition to increased flood protection and reduced flood-insurance rates, Napa residents will benefit from parks and trails for recreation, higher tourism revenues, and improved habitat for fish and wildlife.

Similarly, communities facing increased damage from heavy stormwater runoff can turn impervious surfaces such as roofs, streets, and parking lots into water catchments by strategically planting vegetation. Portland, Oregon, is investing in "green roofs" and "green streets" to prevent sewer overflows into the Willamette River. Chicago, Illinois, now boasts more than 200 green roofs—including atop City Hall—that collectively cover 2.5 million square feet, more than any other U.S. city. The vegetated roofs are helping to catch stormwater, cool the urban environment, and provide space for urban gardens.

Many communities are revitalizing their rivers by tearing down dams that are no longer safe or serving a justifiable purpose. Over the last decade some 430 dams have been removed from U.S. rivers, opening up habitat for fisheries, restoring healthier water flows, improving water quality, and returning aquatic life to rivers. In the ten years since the Edwards Dam was removed from the Kennebec River near Augusta, Maine, populations of sturgeon, Atlantic salmon, and striped bass have returned in astounding numbers, reviving a recreational fishery that adds $65 million annually to the local economy.

## DOING MORE—AND LIVING BETTER— WITH LESS WATER

Of all the water we withdraw worldwide from rivers, lakes, and aquifers, 70 percent is used in agriculture, 20 percent in industries, and 10 percent in cities and towns. With water supplies tightening, we will need roughly a doubling of water productivity by 2025 to satisfy human needs while sustaining nature's life-support systems. Fortunately, opportunities to get more benefit per drop abound through greater investments in conservation, efficiency, recycling, and reuse, as well as through shifts in what is produced where and when.

But the need to do more with less water is not only a challenge for farmers, utilities, and manufacturers. It is also up to individual consumers to shrink our personal water footprints—the amount of water used to produce all the things we buy. The average U.S. resident uses, directly and indirectly, about 2,480 cubic

Table 7.1 Water Used to Produce Selected Products (global average)

| PRODUCT | WATER USED IN PRODUCTION (LITERS) |
| --- | --- |
| 1 tomato | 13 |
| 1 potato | 25 |
| 1 slice of bread | 40 |
| 1 orange/1 glass of orange juice | 50/170 |
| 1 egg | 135 |
| 1 cup of coffee | 140 |
| 1 glass of milk | 200 |
| 1 hamburger | 2,400 |
| 1 cotton t-shirt | 4,100 |
| 1 pair of shoes (bovine leather) | 8,000 |

Source: Adapted from A. K. Chapagain and A. Y. Hoekstra, *Water Footprints of Nations: Volume 1: Main Report*, UNESCO Value of Water Research Report Series 16 (Delft: UNESCO-IHE, 2004), 42.

meters of water per year—about 1,800 gallons *per day*—twice the global average. More conscious choices about what and how much we consume are essential for reducing our global water footprint.

## WATER FOR FOOD

Feeding the world is a very water-intensive enterprise. It takes about 3,000 liters of water to meet a person's daily dietary needs. In the United States, with its high consumption of meat (especially grain-fed beef), the average diet requires some 5,000 liters of water per day. Under some very conservative assumptions, it could take an additional 1,314 billion cubic meters of water per year—equal to the annual flow of 73 Colorado Rivers—to meet the world's dietary needs in 2025.

Once again, the search for solutions needs to begin with a reframing of the question. Instead of asking where we can find 73 Colorado Rivers' worth of water, the question is: *How do we provide healthy diets for 8 billion people without going deeper into water debt?*

Framed this way, the solutions focus on getting more nutritional value per drop of water used in agriculture, which is the key to solving the water-food dilemma (table 7.1).

There are many ways we can grow more food for the world with less water, with most falling into four broad categories: (1) Irrigate more efficiently; (2) boost yields on existing farms, especially rain-fed lands; (3) choose healthy, less water-intensive diets; and (4) use trade to make the smartest use of local water.

### Irrigate More Efficiently

For the last two centuries, societies have focused on expanding irrigation as a key to raising crop production. Today, the 18 percent of cropland that gets irrigation water provides about 40 percent of the world's food—but much of the water withdrawn for farming never benefits a crop. Some of it seeps back into aquifers or nearby streams, while some evaporates back to the atmosphere. There are many ways to reduce the waste: Irrigation can be scheduled to better match crop water needs, for example, or drip irrigation can be used to curb evaporation losses. Reducing irrigation demands by even 10 percent could free up enough water to meet the new urban and industrial demands anticipated for 2025.

### Boost Yields on Rain-Fed Lands

Rain-fed croplands have been the neglected stepchild in global agriculture, but this is now changing. Lands watered only by rain produce 60 percent of the world's food. Some, including those in the U.S. Midwest, achieve very high yields. But many rain-fed farms, particularly in poor countries, produce far less than they could. By one estimate, 75 percent of the world's additional food needs could be met by increasing harvests on low-yield farms to 80 percent of what high-yield farms achieve on comparable land. Most of this potential is in rain-fed areas, and it's achievable through small-scale technologies and improved field methods—including, for example, capturing and storing local rainwater to apply to crops via low-cost irrigation systems. Because the majority of the world's poor and hungry live on rain-fed farms in South Asia and sub-Saharan Africa, raising the farms' productivity would directly boost food security and incomes.

### Choose Less Water-Intensive Diets

Foods vary greatly both in the amount of water they take to produce and in the amount of nutrition they provide—including energy, protein, vitamins, and minerals. It can take five times more water to supply 10 grams of protein from beef than from rice, for example, and nearly twenty times more water to supply 500 calories from beef than from rice. So eating less meat can lighten our dietary water footprint (while also improving our health). If all U.S. residents reduced their consumption of animal products by half, the nation's total dietary water requirement in 2025 would drop by 261 billion cubic meters per year, a savings equal to the annual flow of 14 Colorado Rivers.

## Use Trade to Make the Smartest Use of Local Water

While regional food resilience is important, some water-scarce regions may find it makes better economic and even environmental sense to import more of their food, rather than grow it themselves, and reserve their water for drinking and manufacturing. Egypt, Israel, Jordan, and a dozen other water-scarce countries already import a good share of their grain, saving 1,000–3,000 cubic meters of water for each ton of grain they import. Today, 26 percent of the global grain trade is driven by countries choosing to import water indirectly in the form of grain.

This trade strategy can often be a good alternative to overpumping groundwater or diverting rivers long distances. As water analyst Jing Ma and colleagues point out, northern China annually exports to southern China about 52 billion cubic meters of water indirectly through foodstuffs and other products. This volume exceeds that expected to be shipped from south to north through the massive water-transfer scheme now under construction. A rethinking of where, what, and how food is grown within China might allow the project to be scaled far back, if not eliminated altogether.

At the national level, however, a food policy that relies on grain imports can pose significant risks, especially for poor countries. As China, India, Pakistan, and other populous, water-stressed countries begin to look to the international grain market to meet their rising demands, food prices are bound to increase. The food riots that erupted in Haiti, Senegal, Mauritania, and some half dozen other countries as grain prices climbed in 2007 and 2008 are likely a harbinger of what is to come and suggest that a degree of food self-sufficiency may be crucial to food security. And of course, the rising fuel costs and increased potential for fuel scarcity associated with peak oil will only make food imports more expensive and less reliable in the long run.

## WATER FOR HOMES AND MANUFACTURING

Changes in the production and consumption of manufactured goods can also shrink our water footprints. For example, Unilever is taking steps to reduce water use across the life cycle of its products, from raw materials to manufacturing to packaging to consumer use. Since 1995, water use in its factories has dropped 63 percent, with some of its factories now treating and reusing all of their process water. Unilever is also working with its raw material suppliers to help conserve water. For example, by installing drip irrigation systems on a Tanzanian tea plantation and on a Brazilian tomato farm, the company is shrinking the water footprint of its Lipton tea and Ragu tomato sauce.

In communities across the United States, conservation remains the least expensive and most environmentally sound way of balancing water budgets—and its potential has barely been tapped. Many cities and towns have shown significant reductions in water use through relatively simple measures like repairing leaks in distribution systems, retrofitting homes and businesses with water-efficient fixtures and appliances, and promoting more sensible and efficient outdoor water use. For example, a highly successful conservation program started in Boston in 1987 cut total water demand 43 percent by 2009, bringing water use to a fifty-year low and eliminating the need for a costly diversion project from the Connecticut River.

The greatest residential water-conservation gains yet to be made lie in smarter landscape choices and watering practices. Turf grass covers some 16.4 million hectares (40.5 million acres) in the United States—an area three times larger than any irrigated farm crop in the country. Particularly in the western United States, where outdoor watering typically accounts for 40 to 70 percent of household water use, converting thirsty green lawns into native drought-tolerant landscaping can save a great deal of water. Las Vegas now pays residents $2 for each square foot of grass they rip out, which has helped shrink the city's turf area by 80 million square feet and lower its annual water use by 18 billion gallons in just four years. Albuquerque, New Mexico, has reduced its total water use by 21 percent since 1995 largely through education and by providing rebates to residents for using water-conserving irrigation systems.

One of the biggest untapped potentials for smarter water management in all types of enterprises lies in more creative use of information technologies: meters, sensors, controllers, computers, and even cell phones. A little book-sized product called iStaq, made by U.K.-based Qonnectis, fits under a manhole cover and measures flow, pressure, and other water variables. If the water pipe springs a leak, the iStaq alerts the utility operator by text message. In farming regions, real-time weather data collection combined with crop evapotranspiration rates and sensors monitoring soil moisture are helping farmers determine when and how much to irrigate their crops. There's even an iPhone application that enables farmers to remotely monitor moisture levels in their fields through sensors placed near the roots of their crops.

In Ugandan villages, farmers lacking computers are getting access to the wealth of information on the Internet by calling their questions in to a free telephone hotline called Question Box. The operators, who speak the local language, search for the answers and call the farmers back. A project of Open Mind, a California-based nonprofit, Question Box enables poor farmers, whose only communication device may be a village phone, to connect to the wired world for information on crop prices, weather forecasts, plant diseases, and more.

The potential uses of information technology to enable smarter water decisions are extensive and have only begun to be tapped. Using GIS (geographic information system) technology, for example, the World Wildlife Fund (WWF) recently identified more than 6,000 traditional water tanks (small reservoirs to capture rainfall or runoff) in a single sub-watershed in western India. WWF determined that if the tanks were restored to capture just 15 to 20 percent of local rainfall, they could hold some 1.74 billion cubic meters of water—enough to expand irrigated area in the region by 50 percent and at a cost per hectare just one-fourth that of an irrigation dam-and-diversion project proposed for the region.

## RESETTING THE SIGNALS

Most of the world's water shortages have arisen because the policies and rules that motivate decisions about water have encouraged inefficiency and misallocation rather than conservation and wise use. Without big dams and river diversions subsidized by taxpayers, for example, rivers and streams in the western United States would not be so severely depleted today. And without low, flat rates for electricity, India's groundwater would not be so severely overpumped.

Allowing markets to do what they can do well—send a price signal about water's value—is critical for encouraging investments in water efficiency and more sensible uses of water. Most governments in rich and poor countries alike, however, continue to send the wrong signal by heavily subsidizing water, especially for irrigation, the biggest consumer. While better pricing is essential, it doesn't automatically account for the many important benefits of rivers, lakes, wetlands, and streams—such as protecting water quality and providing fish and wildlife habitat—that are not recognized in the marketplace. It is the job of governments, as custodians of the public trust in water, to protect these important but often unrecognized values, and it is the job of citizens to demand that their elected officials get busy crafting creative solutions.

Imagine, for example, if U.S. policy-makers propped up farm incomes not with irrigation and crop subsidies that distort markets and misallocate resources, but rather with payments for protecting ecosystem services that benefit society at large. Farmers and ranchers who plant buffer strips along streams, protect soils from erosion, or provide wildlife habitat through wetland protection would receive a payment for providing these services. The Conservation Reserve Program under the U.S. Department of Agriculture (USDA) could be strengthened to secure these water benefits for the long term, perhaps in conjunction with the USDA's new Office for Ecosystem Services and Markets. A tax on water depletion or transfers could help fund the effort.

Current pricing and policy signals are deeply misaligned with the realities of our water predicament—but this means that there are untold opportunities for improvement. Each of the ideas listed in box has been implemented by some local, state, or national government

somewhere, and has achieved positive results. For example, a cap on groundwater pumping from the Edwards Aquifer in south-central Texas has motivated farmers, businesses, and citizens to conserve. San Antonio has cut its per capita water use by more than 40 percent to one of the lowest levels of any western U.S. city.

It is critical that policy-makers begin to grapple with the inconvenient truth that supplying water takes energy and supplying energy takes water. Energy and water are tightly entwined, and all too often public policies to "solve" one problem simply make the other one worse. For example, the 2007 mandate of the U.S. Congress to produce 15 billion gallons of corn ethanol a year by 2015 would annually require an estimated 6 trillion liters of additional irrigation water (and even more direct rainfall)—a volume exceeding the annual water withdrawals of the entire state of Iowa. Even solar power creates a demand for water, especially some of the big solar-thermal power plants slated for the sunny Southwest. Clearly any action we take to build local renewable energy sources must be careful not to add additional strain to our already-stressed rivers and aquifers.

The win-win of the water-energy nexus, of course, is that saving water saves energy, and saving energy saves water. The more a community lives on water, energy, and food produced locally, the more options arise for solving multiple problems simultaneously, building resilience through resourcefulness, and preparing for future uncertainties.

## NOTES

1. Intergovernmental Panel on Climate Change (IPCC), *Climate Change 2007—The Physical Science Basis*, (Cambridge, UK: Cambridge University Press, 2007); and IPCC, "Summaries

for Policymakers," in *Climate Change 2007—impacts, Adaptation and Vulnerability*, (Cambridge, UK: Cambridge University Press, 2007).

2  Tim Pearce, ed., "Natural Disasters Displacing Millions—U.N. Study," *Reuters*, September 22, 2009; Robert Draper, "Australia's Dry Run," *National Geographic*, April 2009, 35–59; Heather Timmons, "Half a Million Are Stranded by India Flood," *The New York Times*, September 1, 2008; Mian Ridge, "India's Farmers Struggle Without Crucial Monsoon Rains," *Christian Science Monitor*, August 25, 2009; Paul Rodgers, "Millions Facing Famine in Ethiopia as Rains Fail," *The Independent*, August 30, 2009.

3  "Iowa Flood, Midwest Flooding: Videos, Maps, News and Background," Geology.com, June 13, 2008, http://geology.com/events/iowa-flooding/; "Governor Sonny Perdue Prays for Rain in Georgia, WDEF.com, November 14, 2007; Robbie Brown and Liz Robbins, "Rain Stops, but 8 are Dead in Southeast Floods," *The New York Times*, September 22, 2009.

4  P.C.D. Milly et al., "Stationarity is Dead: Whither Water Management?" *Science* 319 (February 1, 2008), 573–574.

5  Ibid., for $500 billion annual global investment figure.

6  Stuart Bunn, presentation at the 94th Annual Meeting of the Ecological Society of America, Albuquerque, NM, August 2–7, 2009; for an excellent narrative on the Australian drought, see Robert Draper, "Australia's Dry Run," *National Geographic*, April 2009, 35–59.

7  Tim Barnett and David Pierce, "When Will Lake Mead Go Dry?" *Water Resources Research* 44 (March 29, 2008).

8  Richard Seager et al., "Model Projections of an Imminent Transition to a More Arid Climate in Southwestern North America," *Science* 316 (May 25, 2007), 1181–1184.

9  Population figures from Population Reference Bureau (PRB), *2009 World Population Data Sheet*, http://www.prb.org/.

10  Just six countries (Brazil, Russia, Canada, Indonesia, China, and Colombia) account for half the water annually flowing back toward the sea in rivers, streams and underground aquifers—what hydrologists call "runoff," according to the United Nations Food and Agriculture Organization (FAO), *Review of World Water Resources by Country*, (Rome: FAO, 2003).

11.  Number of large dams (those at least 15 meters high) from World Commission On Dams, *Dams and Development* (London: Earthscan Publications, 2000); Jamie Pittock et al., "Interbasin Water Transfers and Water Scarcity in a Changing World—A Solution or a Pipedream?" (Frankfurt: World Wildlife Fund Germany, August 2009).

12  Sandra Postel, "Where Have All the Rivers Gone?" *World Watch 8* (May/June 1995); Fred Pearce, *When the Rivers Run Dry* (Boston: Beacon Press, 2006).

13  Sandra Postel, *Pillar of Sand: Can the Irrigation Miracle Last?* (New York: W.W Norton & Co., 1999).

14  Matthew Rodell, Isabella Velicogna, and James S. Famiglietti, "Satellite-based Estimates of Groundwater Depletion In India," *Nature* 460 (August 20, 2009).

15  V.I. McGuire, *Ground Water Depletion in the High Plains Aquifer, Predevelopment to 2005*, U.S. Geological Survey Fact Sheet 2007–3029, 2007, http://pubs.er.usgs.gov/; 30 percent figure from U.S.G.S., "High Plains Regional Ground Water (HPGW) Study," http://co.water.usgs.gov/nawqa/hpgw/hPgW_home.html.

16  Jamie Pittock et al., "Interbasin Water Transfers and Water Scarcity in a Changing World"; total transfer volume from Ruixiang Zhu, "China's South-North Water Transfer Project and Its Impacts on Economic and Social Development," People's Republic of China Ministry of Water Resources (2008).

17  Kenneth Pomeranz, "The Great Himalayan Watershed: Agrarian Crisis, Mega-Dams and the

Environment," *New Left Review* 58, (July–August 2009); estimates of India's irrigated area vary widely depending on the source and estimation method per Prasad S. Thenkabail, et al., "Irrigated Area Maps and Statistics of India Using Remote Sensing and National Statistics," *Remote Sensing* 1, no. 2 (April 17, 2009), 50–67.

18 Brian D. Richter, "Lost in Development's Shadow: The downstream Human Consequences of Dams," World Commission on Dams, forthcoming.

19 Robert Wilkinson, *Methodology for Analysis of the Energy Intensity of California's Water Systems, and an Assessment of Multiple Potential Benefits Through Integrated Water-Energy Efficiency Measures*, Environmental Studies Program, University of California, Santa Barbara (2000), 6; QEI, Inc., *Electricity Efficiency Through Water Efficiency, Report for the Southern California Edison Company*, (Springfield, NJ: 1992), 23–24.

20 Debbie Cook, former Mayor of Huntington Beach (Calif.), has said "The next worst idea to turning tar sands into synthetic crude is turning ocean water into municipal drinking water.", quoted in "desalination—Energy Down the Drain," The Oil Drum, March 2, 2009, http://www.theoildrum.com/node/5155.

21 Quirin Schiermeier, "Purification with a Pinch of Salt," *Nature* 452, no. 7 (March 20, 2008), 260–261.

22 National Academy of Sciences, Water Science and Technology Board, *Desalination: A National Perspective*, (Washington DC: National Academy Press, 2008); 15,000 figure from Schiermeier, "Purification with a Pinch of Salt."

23 NAS Water Science and Technology Board, *Desalination: A National Perspective*.

24 Sandra Postel, *Liquid Assets: The Critical Need to Safeguard Freshwater Ecosystems*, Worldwatch Paper 170, (Washington, DC: Worldwatch Institute, 2005).

25 Sandra Postel and Barton H. Thompson, Jr., "Watershed Protection: Capturing the Benefits of Nature's Water Supply Services," *Natural Resources Forum* 29, no. 2 (May 2005), 98–108.

26 Cara Goodman, "South America: Creating Water Funds for People and Nature," The Nature Conservancy, http://www.nature.org/. For more examples and a fuller description of Quito's fund, see Postel, *Liquid Assets*.

27 National Research Council, *Valuing Ecosystem Services: Toward Better Environmental Decision-Making* (Washington, D.C.: The National Academy Press, 2005); $366 million figure from David G. Killam, "Sacramento District Project Wins Public Works Project of the Year," website of the U.S. Army, February 12, 2009, http://www.army.mil/.

28 Will Hewes and Kristen Pitts, *Natural Security: How Sustainable Water Strategies are Preparing Communities for a Changing Climate* (Washington, DC: American Rivers, 2009).

29 Emily Pilloton, "Chicago Green Roof Program," Inhabitat, August 1, 2006, http://www.inhabitat.com/2006/08/01/chicago-green-roof-program/.

30 Number of 430 dams is from Rebecca Wodder, "Tolling Bells Ushered in Kennebec River's Rebirth," *Kennebec Journal & Morning Sentinel*, June 28, 2009; $65 million figure is from Hewes and Pitts, *Natural Security*; for more on dams and rivers, see Sandra Postel and Brian Richter, *Rivers for Life: Managing Water for People and Nature*, (Washington, DC: Island Press, 2003).

31 A. Y. Hoekstra and A. K. Chapagain, "Water Footprints of Nations: Water Use by People as a Function of their Consumption Pattern," *Water Resources Management* 21, no. 1 (2006), 35–48.

32 For this calculation, I assumed that the 1.2 billion people who will join humanity's ranks over the next fifteen years will eat low-meat diets, and I made no allowance for the 850 million people who don't have enough food today, nor for the increasing dietary water requirements in China

and elsewhere as incomes rise; hence it is very conservative.

33 Postel, *Pillar of Sand*; 2025 calculation based on withdrawal estimates in William J. Cosgrove and Frank R. Rijsberman, *World Water Vision: Making Water Everybody's Business* (London: Earthscan, 2000).

34 David Molden, ed., *Water for Food, Water for Life: A Comprehensive Assessment of Water Management in Agriculture* (London: Earthscan, and Colombo: International Water Management Institute, 2007).

35 For a fuller description, see Postel, *Pillar of sand*, chapter 9; useful websites include those of the International Water Management Institute, at www.iwmi.cgiar.org, which also has many informative publications; for more on rainwater harvesting, see the website of the Centre for Science and Environment in Delhi, india, at www.cseindia.org, especially their site, www.rainwaterharvesting.org; for examples of affordable, small-plot irrigation, see especially International Development Enterprises, at www.ideorg.org.

36 Dietary water requirement from D. Renault and W.W. Wallender, "Nutritional Water Productivity and diets," *agricultural Water Management* 45 (2000), 275–296; calculation assumes average annual dietary water requirement drops from 1,971 cubic meters per person to 1,242; U.S. 2025 population of 358.7 million is the medium variant estimate of the Population Division of the Department of Economic and Social Affairs of the United Nations, *World Population Prospects: The 2008 Revision*, http://esa.un.org/unpp.

37 Analysis of grain import dependence based on data from U.S. Department of Agriculture, Foreign Agricultural Service, "Production, Supply and Distribution Online," at http://www.fas.usda.gov/psdonline.

38 Jing Ma et al., "Virtual Versus Real Water Transfers Within China," *Philosophical Transactions of the Royal Society B* 361 (2006) 835–842.

39 Postel, *Pillar of Sand*; Sandra Postel, "But Who Will Export Tomorrow's Virtual Water," in *The Truth About Water Wars*, Seed, May 14, 2009, http://seedmagazine.com/.

40 Sandra Postel and Amy Vickers, "Boosting Water Productivity," *State of the World 2004* (Washington, DC: Worldwatch Institute, 2004), 46–65; Unilever example from *Sustainable Development Report 2007: Environmental Sustainability*, (Unilever, 2007).

41 For conservation methods and examples, see Amy Vickers, *Handbook of Water Use and Conservation: Homes, landscapes, Businesses, Industries, Farms* (Amherst, MA: WaterPlow Press, 2001); Boston example from Sandra Postel, *Liquid Assets*, and Sandra Postel, "Lessons from the Field—Boston Conservation," National Geographic website, March, 2010, http://environment.nationalgeographic.com/environment/freshwater/lessons-boston-conservation/.

42 Cristina milesi et al., "Mapping and Modeling the Biogeochemical Cycling of Turf Grasses in the United States," *Environmental Management* 36 (September, 2005), 426–438; Dara Colwell, "Our Love Affair With Our Lawns is Hurling the U.S. Toward Water Crisis," AlterNet, October 2, 2009, http://www.alternet.org/; Las Vegas figures from Robert Glennon, *Unquenchable: America's Water Crisis and What To Do About It* (Washington, DC: Island Press, 2009).

43 Personal email communication with Katherine M. Yuhas, Water Conservation Officer, Albuquerque Bernalillo County Water Authority, Albuquerque, NM, october 12–13, 2009. Between 1995 and 2008, Albuquerque's total water production declined from 40.775 billion gallons to 32.247 billion gallons, while the population served increased from 445,167 to 559,828.

44 iStaq example from Matthew Power, "Peak Water: Aquifers and Rivers are Running Dry. How Three Regions are Coping," *Wired Magazine* 16, no. 5 (April 21, 2008); "iPhone App Offers Remote Water Sensing For Farmers, *The New York Times*, June 30, 2009; see the website for Question Box at http://questionbox.org, and Ron Nixon, "Dialing for Answers Where Web Can't Reach," *The New York Times*, September 28, 2009.

45 Jamie Pittock et al., "Interbasin Water Transfers and Water Scarcity in a Changing World."

46 For more examples, see Sandra Postel and Barton H. Thompson Jr., "Watershed Protection: Capturing the Benefits of Nature's Water Supply Services," *Natural Resources Forum* 29 (May 2005), 98–108.

47 Edwards Aquifer Authority website, at www.edwardsaquifer.org; water use from San Antonio Water System, *2008 Annual Report*.

48 U.S. Congress, Energy *Independence and Security Act of 2007*, 110th Cong., 1st session, 2007.

49 R. Dominguez-Faus et al., "The Water Footprint of Biofuels: A drink or Drive Issue?" *Environmental Science & Technology* 43 (May 1, 2009), 3005–3010.

50 Todd Woody, "Alternative Energy Projects Stumble on a Need for Water," *The New York Times*, September 30, 2009.

# WHY DO WE USE SO MUCH ENERGY, AND WHAT FOR?

By Fereidoon P. Sioshansi

## INTRODUCTION

When asked if he wanted India to attain a living standard similar to that enjoyed by Britain, Gandhi reportedly replied—I am not sure about his exact words—that Britain had attained its high standard of living by exploiting the resources of half the world, a reference to its colonies. He then asked, "How many worlds do you think India will need to reach a similar living standard?"[1]

To be sure, a significant portion of India's population enjoys much higher living standards today and we have not run out of natural resources yet. One can also excuse Gandhi's politically loaded response in the context of British colonial policies during India's long struggle for independence. But his observation about our ever-increasing reliance on finite natural resources to provide a decent standard of living for people of India—or for that matter, the world—is as relevant today as when he made the statement. Moreover, Gandhi's remarkable remark goes to the core of the central question, namely can we maintain an adequate standard of living in a sustainable fashion for the world's growing population—not just for the rich and the privileged?

Different scholars come up with different answers to Gandhi's question. Some believe that continued technological advances will allow us to meet the ever-increasing demands of a growing population (Figure 1), no problem. Others, while acknowledging the significant role of technological fixes, are not so sure.

Aside from long-term sustainability is the concern for the lack of equity and fairness in the current distribution and use of resources. Today, we live in a world where an estimated one-third of the global grain harvest is used to feed the animals that produce the meat that is consumed by the rich, while millions of people in less fortunate circumstances struggle to feed

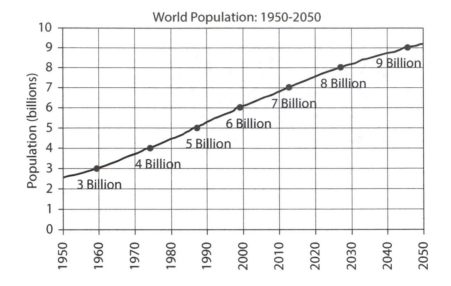

FIGURE 1. WORLD POPULATION PROJECTION TO 2050.

Source: US Census Bureau.

themselves, lack clean water, electricity, or other basic necessities.[2]

It is in this context that this chapter explores a number of critical issues affecting the future course of human evolution on planet Earth.

This chapter, benefiting from the insights of the others and referring to them selectively, is organized as follows:

- Section 2 asks *why* we use so much energy.[3]
- Section 3 briefly examines *what* we use energy for.
- Section 4 asks what constitutes an adequate standard of living and how much energy we need to maintain it.
- Section 5 examines alternative lifestyles, habits, and socio-economic systems that might provide equally satisfying—or potentially superior—standards of living on a fraction of our current levels of energy consumption.
- Section 6 asks what would it take—in terms of technological improvements, changes in policies, socio-economic

systems, regulations, education, cultural, and behavioral adjustments—to lead us toward a more sustainable future path.

This chapter is primarily focused on energy use and its environmental consequences, but one can easily substitute the term *natural resources* for *energy*. By the same token greenhouse gas (GHG) emissions are associated with burning of fossil fuels (Figure 2) and resource depletion issues—visible symptoms of our current unsustainable practices.

The chapter's main contribution is to encourage a fundamental search for alternatives to the *status quo* by asking questions such as, why can't things be radically different in the future than they have been in the past? Or, why can't we accomplish far more with far less?

## WHY DO WE USE SO MUCH ENERGY?

The short answer is that *up to now* energy has been relatively cheap and seemingly plentiful

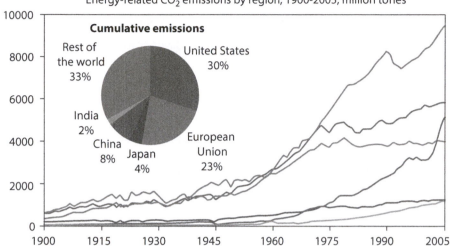

**Keeps rising**

Energy-related $CO_2$ emissions by region, 1900-2005, million tones

FIGURE 2. ENERGY-RELATED $CO_2$ EMISSIONS.

Source: International Energy Agency, *World Energy Outlook*, 2007.

and few were concerned about climate change, sustain-ability, or equity issues (Figure 3).

Economic theory predicts that, all else being equal, one would use more of one critical input in the production process if it were inexpensive relative to others. If energy were cheap relative to labor and capital—typical variables in the production process—the rational user would substitute more energy while cutting back on labor and/or capital. This has been a recurring pattern since the beginning of the industrial revolution, as illustrated in the following anecdotes:

- The longevity of the notoriously inefficient incandescent light bulb, virtually unchanged since Thomas Edison invented it a century ago, provides a manifestation of this phenomenon. These bulbs typically *waste* over 90% of the energy in the form of unwanted heat,[4] converting less than 10% of the energy into useful light. Yet the incandescent light bulb is still in widespread use because electricity has traditionally been relatively cheap, subsidized, or inappropriately priced.[5]

- Until recently, computer makers paid scant attention to energy use, focusing exclusively on other features such as speed, capacity, and other product attributes. All that was needed to get rid of the generated heat was a bigger fan—or in the case of large main-frame computers, more air-conditioning—at the expense of the users, not the manufacturers.[6] It was not until portable computers became popular that computer makers began to pay serious attention to power consumption, primarily driven by limited capacity of batteries, especially in early years.

- Detroit's big three carmakers, all suffering financially today, for decades paid virtually no attention to fuel economy, focusing entirely on other features—including making progressively bigger, heavier, and more gas-guzzling models.

- Planes, trains, buses, trucks, ships—and virtually all other major energy consuming devices—were and still are designed and marketed on features other than energy efficiency, be it speed, capacity, range,

## What Happens When Fuel Is Cheap and Emissions Are Free?

It was recently reported that commercial transatlantic flights can reduce their fuel consumption by roughly 2% by making trivial modifications in operating procedures such as flying at optimal altitudes for maximum efficiency rather than arbitrary altitudes set by air traffic controllers. This requires *no* technological improvements and virtually *no* investment, in other words *zero pain, 2% gain.*

Given the current traffic levels in the key transatlantic market—100,000 oneway flights annually, each one on average using 25 metric tons of fuel—a 2% saving adds up quickly. Similar savings can be had on other routes and other markets. Less fuel burned means less GHG emissions, which are especially damaging when released at high altitudes. The question is why such trivial measures were not explored until now?

The answer is that airlines began to pay serious attention to their fuel consumption only after recent rises in the cost of jet fuel, beginning in 2008 when oil prices approached nearly $150 per barrel for the first time. They became even more interested in fuel consumption only after the European Commission began talking about putting restrictions on GHG emissions associated with air transportation, believed to represent roughly 3% of global GHG emissions.

Two observations:

- First, if energy costs are a trivial part of the overall cost of a business, they get scant attention.
- Second, regulations—or the mere threat of regulations—are usually needed to prompt the industry to take corrective measures.

As further explored, the implications of this example are twofold.

- First, prices must reflect the true and full costs. To the extent that they do not capture externality costs—in this example GHG emissions associated with planes at high altitudes—they lead to suboptimal outcomes.
- Second, regulations, standards, and policy imperatives are often needed to correct market imperfections, which are typically driven by short-term profit motives. The question of how much regulation and how intrusive it should be, however, is not as trivial as it may seem.

**Source**: *Atlantic Interoperability Initiative to Reduce Emissions (AIRE), reported in* The Wall Street Journal, *10 March 2010.*

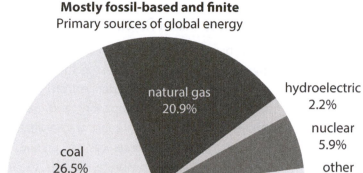

**Mostly fossil-based and finite**
Primary sources of global energy

FIGURE 3. WORLD PRIMARY ENERGY SOURCES.

Source: R. Heinberg, *Searching for a Miracle*, 2009, based on EIA Data.

performance, longevity, maintenance, and most important, initial price tag. If a plane's initial price is less than a rival model or if has a longer range, *that* becomes the main selling point—not its fuel consumption, which often dwarfs the initial price tag. Fuel consumption has become a selling point only in recent times due to rising fuel prices and concerns about pending regulations (see "What Happens When Fuel Is Cheap and Emissions Are Free" sidebar).

As these examples illustrate, the pattern is pervasive.[7] It is not limited to a few products, few industries, or a few countries. Even today, manufacturers of most consumer products focus almost exclusively on minimizing the initial price of the finished product—using material and designs that reduce manufacturing costs even when this increases life cycle energy consumption since these costs are borne by consumers.[8] This raises important questions about *whether* and *how* to protect consumers who may not necessarily be well-informed and/ or sufficiently motivated to compare initial purchase price of a product to its lifelong energy consumption costs.[9]

Over time, the *illusion* of cheap and seemingly abundant energy[10] has instilled wasteful habits in many of us—and these habits have permeated throughout our lives, lifestyles, livelihood, infrastructure, and institutions to the point were we are no longer even aware of them. People in rich countries have acquired energy-thirsty lifestyles including long commutes to work, for example. In many cases, prevailing customs, social norms, and even our tax codes actually *favor* increased consumption[11] (see "Do You Want to Start a Revolution" sidebar).

And because energy has historically been cheap and seemingly plentiful, no one thinks twice about traveling to a distant city for a sales call or a weekend skiing holiday. Dubai boasts a giant indoor ski resort, which is kept at freezing

## Do You Want to Start a Revolution?

During a recent trip to Germany, I was picked up by a friend in a Mercedes Benz. I could not resist congratulating him on owning such a fine car. He confessed that it was a *company car*. Further questioning revealed that he commuted a long distance from home to work, and beyond, on a daily basis. When asked about the cost of gas, parking, and maintenance, he said everything was covered by his employer. Why bother with public transportation, he implied?

Noticing how puzzled I was, he explained that his case was not unique, that thousands of German mid- to top-level managers enjoy similar corporate benefits, the higher the rank, the more prestigious the car. In fact, he said many of his colleagues would prefer a fancier car to higher pay. More than your salary or title, the brand of the car and the size of the engine determines your ranking and status within the company, he explained. And, of course, the same goes for the location of the assigned parking space in the corporate garage.

When I pointed out that this encourages more driving, more traffic congestion, more gas consumption, more pollution—all *negative* externalities associated with driving cars—he looked at me as if I had come from Mars. He asked, "What are you trying to do, start a revolution?" More company cars, he explained, means more jobs for the likes of Mercedes Benz and the entire value chain that produces the parts and components. Driving cars generates additional jobs for the petrol industry, for road construction and maintenance, not to mention generating tax revenues to pay for other services. Plus all the other *positive* benefits—jobs for the tire industry, car service industry, insurance industry, and so on.

This, and numerous examples like this, shows the pervasive consumption and production culture we live in. My conversation with my friend made me aware of the myriad customs and conventions that lead to *more* consumption, because it creates demand for *more* production, and that is what keeps the economic engine moving—but to what end?

temperatures in a hot climate, not to mention the glittering lights of Las Vegas and numerous other examples of gluttonous and wasteful energy use. How can anyone seriously begin to discuss energy efficiency in Las Vegas, Abu Dhabi, or Houston?

The rapidly growing ranks of the well-off of the world can afford their huge carbon-footprint because energy is a small percentage of their disposable income. The popularity of SUVs, second holiday homes, and remote resorts in exotic places are symptoms of the energy glut supported through hidden subsidies, convoluted tax laws, and other perverse incentives that encourage consumption as opposed to conservation.

Making matters worse, in many countries, energy continues to be heavily subsidized, resulting in excessive consumption (see "Promoting Wasteful Consumption Through Energy Subsidies" sidebar). Is it any surprise that citizens of many oil-rich countries in the Persian Gulf, who enjoy subsidized petrol at below market prices aimlessly roam around in gas-guzzling SUVs without a second thought (Figure 4)? The same is true of other precious necessities, notably water and food, which are subsidized in many parts of the world, resulting in excessive waste and inappropriate use.[12]

Surprisingly, energy subsidies are not limited to developing countries with vast domestic energy resources. Many net importers of energy (e.g., India and China) subsidize energy prices, in some cases literally bankrupting their economies in the process (e.g., Iran and Venezuela) while others are projected to become net importers of oil at current growth rates (e.g., Malaysia and Egypt). The total amount of subsidies outside the Organisation for Economic Co-operation and Development (OECD) countries is estimated at a staggering $310 billion a year.

Nor is the practice limited to developing countries. The U.S. fossil fuel industry—oil, gas, and coal—reportedly received $72 billion in subsidies between 2002 and 2008, according to the Environmental Law Institute. The nuclear industry gets significant subsidies also—no one can be sure exactly how much. More recently, renewable energy resources have been receiving subsidies. Their proponents consider these more deserving than subsidies given to fossil fuels on environmental grounds as well as for energy security and fuel diversity reasons.

We not only *use* a lot of energy but we also *waste* a lot of it unnecessarily. According to one estimate, as much as a third of energy used in the U.S. commercial sector may be *unnecessarily* consumed or wasted.[13] And this explains why much of our infrastructure is poorly designed, our cities and urban centers have turned into endless sprawls, and our personal habits and lifestyles have become energy intensive.

Another issue, of course, is not just *how much* energy but *what kind* of energy? For a long list of reasons, we have become overly dependent on fossil fuels, with significant implications for the environment.[14]

## WHAT DO WE USE ENERGY FOR?

The short answer is that we use energy to *derive* needed services.

A lump of coal, a barrel of oil, a kilowatt-hour of electricity, or a cubic feet of natural gas are only valued because they can heat a furnace, run a car, allow us to watch TV, or fry an egg, respectively. This is referred to as *energy services;* we need the *services* provided by energy, not the energy source itself. The petrol in the car tank contains refined and concentrated energy

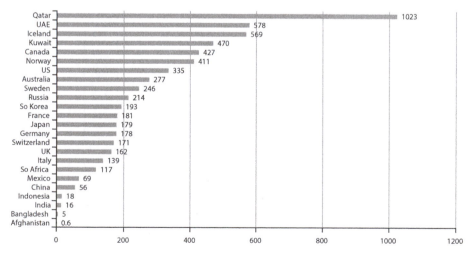

FIGURE 4. WHAT ABOUT THE INEQUALITIES? PER CAPITA ENERGY USE IN SELECTED COUNTRIES, IN MILLION BTUs/CAP, 2006 DATA

Source: *Searching for a Miracle,* based on EIA data.

## Promoting Wasteful Consumption Through Energy Subsidies

Energy subsidies, if they ever made sense, were justified on the grounds that they help the poor. And some forms of subsidy, say for bread or mass transit, could conceivably be justified in some cases and they may in fact help the poor. But energy price subsidies hardly ever qualify.

Since the middle class and the rich use far more energy than the poor—they drive bigger cars, own bigger homes, have bigger air conditioners, fly more often, and so on—they benefit disproportionately from energy price subsidies, be it petrol prices, electricity, or natural gas. It is hard to imagine the opposite. In this sense, energy price subsidies are among the most distasteful forms of regressive tax imaginable.

But that is not the end of it. Artificially subsidized energy prices encourage increased consumption, which runs counter to the goal of encouraging energy conservation and more efficient energy utilization. Subsidies also empty government coffers, money that could be put to better use elsewhere.

And of course, more consumption results in more carbon emissions. At the G20 Summit in Pittsburgh in 2009, leaders of the 19 biggest economies plus the European Union agreed to *phase out energy price subsidies* in *medium term*, at least in principle. This is expected to result in a 10% drop in global GHG emissions by 2050.

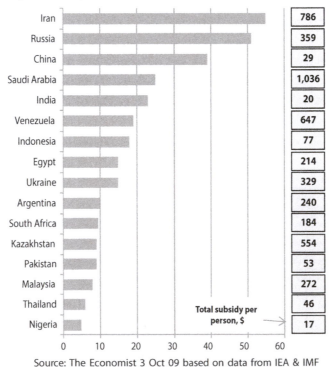

**Your subsidy is dumber than mine**
Energy subsidies among selected non-EOEC countries, 2007 data, $billion and in $ per capita

| Country | Subsidy ($billion) | Total subsidy per person, $ |
|---|---|---|
| Iran | ~56 | 786 |
| Russia | ~52 | 359 |
| China | ~40 | 29 |
| Saudi Arabia | ~25 | 1,036 |
| India | ~23 | 20 |
| Venezuela | ~19 | 647 |
| Indonesia | ~18 | 77 |
| Egypt | ~15 | 214 |
| Ukraine | ~15 | 329 |
| Argentina | ~10 | 240 |
| South Africa | ~9 | 184 |
| Kazakhstan | ~9 | 554 |
| Pakistan | ~9 | 53 |
| Malaysia | ~8 | 272 |
| Thailand | ~6 | 46 |
| Nigeria | ~5 | 17 |

Source: The Economist 3 Oct 09 based on data from IEA & IMF

that allows us to drive a long distance without refueling and that provides a highly valued service: mobility. The kilowatt-hours that keep the lights on, allow us to listen to music, or access the Internet, provide lighting, entertainment, and connectivity, respectively—services we greatly value and desire.

Even though the concept of *energy services* gets us one step closer to answering the question, "What do we use energy for?", it only goes so far. For example, consider the need for mobility. Depending on the distance and the allotted time, there are many ways to get from point A to point B with significant cost, energy, and carbon implications. Assuming a modest distance, one can bicycle, walk, take a bus, share a ride, or drive a private car. In this case, the choices are listed in terms of increased levels of energy consumption. There are, of course, other important attributes associated with these choices, notably time, comfort, safety, style,[15] and so on.

To make the example more complicated, the car can be large and heavy or small and light requiring different amounts of energy to provide virtually the same service. Depending on the value placed on energy, time, convenience, and other factors—perhaps it is cold, hot, or raining—one option may turn out to be superior to one person while another suits a second.

The ultimate choice of what is the best way to get from A to B, however, can be *influenced* by many factors. For example:

- If a free shuttle bus runs frequently between point A and B, many may be persuaded to take the bus.
- Having safe designated bicycle lanes between A and B and parking racks at both ends may persuade some people to ride the bike.[16]
- Availability and price of parking is often a significant determinant of choice. If parking is scarce and expensive, other options become more attractive.[17]
- Road congestion is often an important factor. Some countries offer multiple-occupancy lanes, allowing cars with two or more riders to use specially designated lanes as a way to promote high occupancy in private cars.[18]
- A pleasant pedestrian walkway between A and B may encourage more people to consider walking relatively short distances, especially in congested urban areas.
- Prices, including fuel tax, bus fare, and parking fees, and also road access restrictions will influence the choice.[19]

Why dwell on such a trivial example? The point is that urban planners and governments can influence the final choice of travel mode, say, away from private cars to other options with multiple benefits including less congestion, less energy consumption, less pollution, and less space allocated to roads and car parking, leaving more to city parks and open space. Many cities are promoting improved mobility and access, and less congestion and pollution, by increasing urban density through better planning.[20]

Private cars, considered by many as the ultimate epitome of comfort, personal convenience, and flexibility will be hard to replace by buses, bikes, or other means of transportation, but much can be done to reduce their energy consumption *and* emissions—the most troubling source of pollution in urban areas.[21] The good news is that today's cars, like incandescent light bulbs, are energy guzzling monsters compared to what they can become.

Depending on many variables, today roughly 90% of the energy content of the fuel in a typical internal combustion (IC) engine car may be wasted as heat—not delivered as mobility. One cursory look at an IC engine reveals the ingenious engineering that goes into capturing

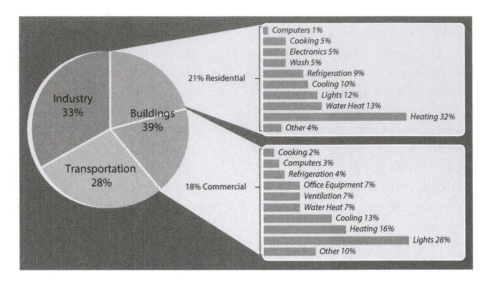

FIGURE 5. WHY SO MUCH ENERGY? US ENERGY CONSUMPTION BY MAJOR SECTOR.

Source: Lawrence Berkeley National Laboratory.

and disposing of enormous quantities of heat into the atmosphere via the radiator, the hoses, and the fans. By comparison, an electrical engine uses roughly 90% of the energy to provide mobility with no emissions at the point of use—which explains the current interest in electric and hybrid vehicles.[22]

The most energy frugal, zero congestion, zero emission mobility option, of course, is to avoid the need for *physically* going from point A to B altogether. Perhaps we can conduct the required business on the phone or by accessing the Internet or by sending a text message. Examples include remote banking, remote shopping, and telecommuting. An increasing number of people now routinely work from a home-based office or telecommute,[23] with considerable implications for energy use and GHG emissions while saving time and avoiding the stress of daily commuting.[24]

Figure 5 provides some clues to the question of what we use so much energy for. The mobility example above can be repeated in countless other contexts. Take the case of lighting, a significant contributor to energy use in both residential and commercial sectors.[25] In

many cases, artificial lighting can be significantly reduced by increased reliance on natural lighting during daylight hours. This, of course, requires better design, building orientation, windows, shades, and other means to adjust interior lighting without increasing the buildings' heating load.

Artificial lighting can be provided by more or less efficient lights and lighting fixtures. For example compact fluorescent lights (CFLs) use a quarter of energy per lumen than traditional incandescent light bulbs. And light emitting diodes (LEDs) offer even more drastic energy savings—albeit at a cost.

But the cost is not fixed. It can be *influenced* by supportive policy measures, standards, and regulations. CFLs used to be pricey and their performance and light quality was poor when first introduced.[26] CFL prices have plummeted in the recent past as a mass market has evolved, and prices continue to decline while quality and performance improves.[27] The same can hopefully be said about LEDs, and the next generation of super-efficient lighting technologies and fixtures.

Moreover, there is more attention devoted to how the artificial light is directed to where it is needed. Directional lighting and more efficient lighting fixtures deliver a large percentage of the light to where it is needed, which means little energy is wasted to illuminate space that does not need the lumens. The savings from such simple applications, aggregated over a large population of users, can be enormous.

There is, of course, an extremely low-tech, zero-cost option to cut down energy used for lighting to zero, a parallel to the case of telecommuting—the option that literally negates the need for mobility in the prior example. It is turning off the lights when not needed, or not turning them on in the first place.

Other examples of "What do we use so much energy for?" can come from the commercial and industrial sector.[28] In many commercial establishments, huge amounts of energy are wasted; for example, the air conditioners work overtime to keep the premises cool while doors and windows remain open. In many restaurants, the oven and various heating devices used to cook food generate enormous amounts of heat that the air conditioner and the circulation system have to get rid of. Better, more integrated designs can overcome many of these types of wasteful energy consumption or recycle the wasted heat into a useful application.

The commercial sector, which captures everything that is not residential, industrial, or agricultural, poses special challenges when it comes to energy use because the typical users/occupants do not generally pay for the energy consumed directly, nor do they have direct incentives to conserve. Moreover, energy costs tend to be a relatively small percentage of overall costs of products and services in the commercial sector, making energy efficiency a difficult proposition in many organizations.[29]

Typical thermal power plants generate enormous heat that must be exhausted, at considerable cost and harm to the environment. Combined heat-and power, cogeneration, and district heating provide useful applications for the wasted heat.[30]

The average homes in many countries are poorly insulated, allowing too much heat to get in during the summer, when it is not needed, and too much heat to escape from the interior during the winter, when it is. Making matters worse, generations of architects and engineers were taught to treat buildings as isolated spaces to be artificially heated, cooled, lighted, humidified, and ventilated with little consideration of the ambient environment. This thinking is now being challenged, and changed, in favor of designs that *integrate* the building into the surrounding environment, taking advantage of the orientation, natural lighting, and the ambient air to meet most of the comfort needs of the occupants.[31]

In sum, we typically *waste* a lot of energy in the process of getting to what we really want and need. It is perhaps a slight exaggeration to ask, "Do we need a tank to kill a fly?"

## WHAT CONSTITUTES AN ADEQUATE STANDARD OF LIVING?

This is an important question and highly relevant if we agree, as the prior discussion suggests, that when it comes to energy consumption, we, in fact, often use a tank to kill a fly, rather than a fly swatter.

If we can define an *adequate* standard of living, and identify how much energy services are needed to support that level of comfort, then we can focus on ways to provide the services with the least amount of energy.[32]

By most measures, people living in rich countries, on average, enjoy high standards of living made possible by prodigious use of energy and other resources (Figure 6).[33] Referring to Gandhi's observation at the beginning of the chapter, these high living standards are made possible by reliance on cheap imports of natural resources and, increasingly, manufactured products, from developing countries.[34]

In contrast, many people in developing countries subsist on the mere necessities on a fraction of energy and natural resources used by citizens of developed countries. The average per capita energy use and carbon emissions of the former are miniscule compared to the latter group (Figure 7).[35]

Defining what constitutes an adequate standard of living, however, is fraught with difficulties:

- First, standard of living for whom?
- Second, even for a given population, for some of the reasons described in preceding sections, it is not trivial to define how much energy and natural resources are *needed* to sustain a given standard of living.

The first problem goes back to the immense inequalities that have arisen among various societies and cultures over the course of human evolution and economic history.[36] As further explained by Bartiaux et al. (Chapter 3), different societies have different definitions for what constitutes basic needs.[37] In fact, these authors suggest that *energy needs* may not provide a useful basis for defining *energy demand* or arriving at future energy projections. More fundamentally, we are faced with the reality that the definition of human need is problematic on so many dimensions as to be virtually indefinable (see "How Many Gallons Does It Take for a Shower?" sidebar).

So where does one draw the line between personal choice and societal costs—in this case waste of energy and water in a finite planet? Where does basic human *needs*, such as cleanliness, end and where does frivolous consumption begin? Clearly, the question is not limited to use of water and goes beyond the definition of a showerhead. The point of this anecdote is to illustrate the elastic nature of need, in this case water for taking a shower.

Swiss citizens might define an adequate standard of living based on what they have come to expect as reasonable and customary.[38] The same question posed to citizens of Swaziland may produce a rather different answer. The difference in the two answers can be attributed to different standards of living in the two countries, as well as other variables including population density, climate, culture, lifestyles, energy prices, income levels, and so on.

But it gets even more complicated than that (see "How to Measure Prosperity or Happiness" sidebar). Currently, a typical Swiss citizen uses roughly half of the energy of a typical American on a per capita basis. Most people would agree that we cannot therefore conclude that a typical American enjoys a standard of living twice that of a typical Swiss. In fact, by some measures, standards of living may be higher in Switzerland than in the U.S.[39] Clearly other factors play a role. Switzerland is a small, mountainous, landlocked country with an extensive train and mass-transit system that allows many citizens to enjoy comfortable lives without the need for an automobile, let alone a big SUV. The typical Swiss lives in a smaller house, with smaller appliances, drives a smaller car fewer miles—if at all—and far less often than the typical American.

But the issue of how much energy is needed to sustain an adequate lifestyle is even more complicated than comparing Swiss to American lifestyles. Looking at per capita energy consumption and carbon emissions among selected U.S. states, what explains the vast difference between Wyoming and California (Figure 8)? Retail electricity prices, population density, the composition of the economy, climate, home-building codes, and appliance standards explain some of the difference but not all. In the case of Wyoming, a mining and coal producing state with a small population, the per capita number

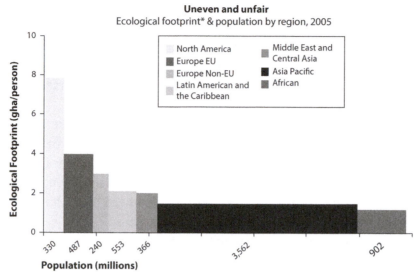

**Uneven and unfair**
Ecological footprint* & population by region, 2005

FIGURE 6. ECOLOGICAL FOOTPRINT.

Source: *Living Planet* 2008.

looks excessively high. California, on the other hand, is a state with a relatively mild climate, little heavy industry, and high retail electricity prices.

Even if one adjusts for these variables, defining how much energy—and other natural resources—is needed to sustain an *adequate* standard of living is still fraught with challenges. The reason, as already explained, is that our basic *energy service needs* can be met through a combination of energy, capital, and human factors. By adjusting these variables, one can arrive at vastly different answers. Pushing the concept to its logical limits, one can approach *zero net energy use* in buildings if costs were not an obstacle and one could invest not in the *leading edge*, but as Amory Lovins likes to say, *bleeding edge* technologies.[40]

To demonstrate the point, let's examine the energy use of a building. A highly insulated building with state-of-the-art windows, lighting, and space conditioning equipment and controlled with a sophisticated energy monitoring and management system would use very little energy, offering significant cost savings over its

extended life.[41] But such a building comes at a relatively high up-front cost.[42] A *zero net energy home*, will—at least in theory—have a *zero* net energy consumption over time but to get there would require a significant investment in design and construction, not to mention highly efficient energy using and generating devices and behavioral adjustments by its occupants.[43]

Since there are obvious tradeoffs between how much is invested in the capital stock versus their energy usage over time, this can quickly turn into a circular argument.[44] From an economic point of view, energy efficiency investments should be pursued to the point where they are deemed economic in terms of reduced operating costs. This entails putting a value on future energy savings versus present investments in better equipment, better design and construction, more insulation, and so on. Since future savings are worth less than present investments, consumers must make tradeoffs, explicitly or implicitly, when buying energy-intensive appliances, cars, or homes. A number of writers examine the implications of

## How Many Gallons Does It Take for a Shower?

Water consumption is energy-intensive. It must be collected, pumped, stored, purified, disinfected, and piped. After usage, it must be collected, pumped, treated, and discharged—all of which requires energy. Not to mention that it is becoming increasingly scarce in many parts of the world, requiring highly energy intensive desalination of sea water.

For affluent consumers, none of this matters. In the last few years, there has been a trend toward multiple nozzles and massive showerheads that discharge enormous quantities of water. Moreover, consumers in rich countries take more frequent and longer showers. Students residing in U.S. college dormitories average 14-minute showers, and counting. Many luxury spas now advertise the fact that you can enjoy an 18-showerhead bath with incredible water flow. Obviously no one is going to time how long a shower takes.

Given water scarcity and its energy intensity, water flow, pressure, and the definition of showerhead becomes critical. As it happens, a 1992 U.S. federal law specifies that a *showerhead* cannot deliver more than 2.5 gallons per minute (GPM) at a pressure of 80 pounds per square inch (PSI). Some current showerheads in the U.S. deliver as much as five times this limit. Making matters worse, an increasing number of upscale new homes—industry estimates put the number in the 1%–4% range and growing–now feature showers with multiple showerheads, as many as 18, spraying water from all directions.

The U.S. Department of Energy (DOE) has decided that enough is enough. In May 2010, according to an article in *The Wall Street Journal* (21 July 2010), Scott Harris, DOE's General Counsel, fined four shower manufacturers for failure to abide by the law, putting others on notice. Moreover, the DOE has decided that the definition of a showerhead needs further elaboration.

## Watch Out for Water Police

Water usage for a 3-minute shower with a showerhead compliant with 1992 U.S. federal law and a 15-minute shower using 18-nozzle showerheads, in gallons:

| Current U.S. law | Luxury 18-nozzle showerhead |
|---|---|
| 3-minute shower | 15-minute shower |
| 7.5 gallons | 675 gallons |

Manufacturers' interpretation of the existing law is that as long as *each* showerhead complies with the existing mandate, there is no limit to how many *nozzles* there can be in a shower. Hence, a bathroom with 18 nozzles each delivering 2.5 GPM, or 45 GPM in total, would be legitimate. In this case, a 15-minute shower will use 675 gallons of water, more than people

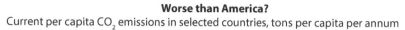

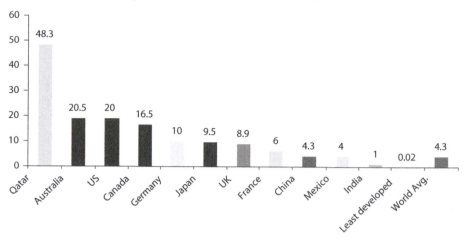

FIGURE 7. WORSE THAN AMERICA? CURRENT PER CAPITA CO₂ EMISSIONS IN SELECTED COUNTRIES, TONS PER CAPITA PER ANNUM.

*Energy Informet,* Oct 2009.

## How Many Gallons Does It Take for a Shower?—cont'd

in some impoverished countries use in a whole year. DOE's interpretation, however, is that all nozzles would count as a single showerhead and subject to the 2.5 GPM limit. That would mean that if you choose to have 18 nozzles, each would deliver a trickle, not a flood of water.

This has caused quite a fury among the manufacturers. Barbara Higgens, Executive Director of the Plumbing Manufacturers Institute—yes, apparently there is such an institute—complained to the *Wall Street Journal* that DOE's Mr. Harris is making a "value judgement," adding that, "One person's waste is another person's therapeutic use of water." Pedro Mier, Vice President of Grupo Helvex in Mexico, agreed, pointing out that his firm's customers "just like to feel they are getting a lot of water." He confessed that until he received the DOE's letter recently, he was not even aware of the federal law. Referring to Mr. Harris' letter, he said, "At first, I thought it was a scam."

"Did Congress limit consumer choice?" DOE's Mr. Harris asks rhetorically, adding, "Absolutely. When you waste *water*, you waste *energy*," (emphasis added) and that's where DOE comes into the picture. Mr. Harris estimates that each multihead shower fixture uses the equivalent of roughly one barrel of oil per year. It may not sound like a lot, but multiplied over many homes built each year, it quickly adds up.

these intertwined issues and how appropriate policies, incentives, information, standards, and codes can *influence* the ultimate decision.

This section must end inconclusively on both questions posed, namely,

- There is *no* universally accepted definition of what constitutes an adequate standard of living,
- There is *no* definite answer to how much energy—and other resources—it takes to sustain it even if a universal definition existed.

For pragmatic purposes, however, one can assume that living standards enjoyed by Western European countries today are adequate, and average Europeans appear to get by on half as much energy per capita than their American, Canadian, or Australian counterparts. In the case of Switzerland, a country with enviable living standards, there is an effort to cut the current per capita energy use to a third of current levels.

Switzerland's ambitious target may provide a useful goalpost, say, the equivalent of 2000 Watts, or even less, per person. To put this number in perspective, the corresponding number for the U.S. is currently around 12,000 Watts per person—implying that the U.S. would have to cut its per capita energy consumption six times as much to achieve what the Swiss have established as a goal. This, however, is a gross simplification given the vast differences in the composition of the economies, population density, and numerous other factors.

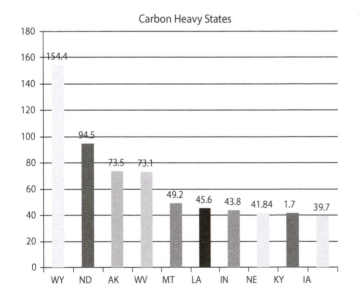

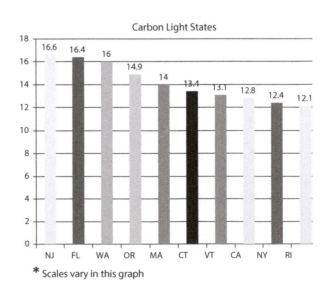

FIGURE 8. UNITED, BUT NOT EQUAL, STATES OF AMERICA. TOP 10 CARBON-INTENSIVE AND CARBON-LIGHT STATES IN THE US, PER CAPITA GREENHOUSE GAS EMISSIONS IN TONS OF $CO_2$ EQUIVALENT PER.

Source: Climate Analysis Indicators Database, World Resources Institute.

## How to Measure Prosperity or Happiness?

Economists traditionally measured a nation's prosperity based on its economic output, per capita GDP, or similar monetary indexes. But as everyone knows, money alone doesn't buy happiness. The debate about how to measure prosperity, and happiness, has received more attention in recent years with suggestions that nonmonetary factors, such as quality of a nation's institutions, the state of its democracy, environment, its health, education, public services, income distribution, freedom, security, and leisure time play important roles.

French President Nicolas Sarkozy recently commissioned a study led by prominent economist Joseph Stiglitz to come up with alternative measures, pointing out that a nation's GDP is an "insufficient measure of its wellbeing." Few would disagree.

Legatum, a London-based think tank has come up with a multidimensional prosperity index, which includes a number of nonmonetary measures. It ranks Finland on top, followed by Switzerland, Sweden, Denmark, and Norway—all reasonably wealthy countries but also enjoying other positive attributes such as stable democratic institutions, good social welfare systems, reasonable distribution of income, among other reasons.

The U.S. is ranked ninth, ahead of Britain, Germany, and France. Zimbabwe, Sudan, and Yemen did not do well—not only because they are poor but, more importantly, because they suffer from unstable and undemocratic governments, if one can dignify their ruling system as a government.

| Overall Rank | Country | Economic Fundamentals | Entrepreneurship & Innovation | Democratic Institutions | Education | Health | Safety & Security | Governance | Personal Freedom | Social Capital |
|---|---|---|---|---|---|---|---|---|---|---|
| 1 | Finland | 10 | 9 | 9 | 3 | 7 | 2 | 2 | 7 | 6 |
| 2 | Switzerland | 2 | 13 | 1 | 22 | 3 | 6 | 3 | 11 | 2 |
| 3 | Sweden | 16 | 3 | 7 | 4 | 15 | 7 | 5 | 5 | 3 |
| 4 | Denmark | 15 | 6 | 12 | 2 | 12 | 4 | 1 | 2 | 13 |
| 5 | Norway | 18 | 17 | 8 | 1 | 10 | 1 | 7 | 1 | 10 |
| 6 | Australia | 7 | 15 | 5 | 6 | 21 | 14 | 10 | 4 | 4 |
| 7 | Canada | 6 | 4 | 6 | 16 | 22 | 9 | 9 | 3 | 9 |
| 8 | Netherlands | 3 | 5 | 19 | 14 | 8 | 15 | 8 | 10 | 8 |
| 9 | United States | 14 | 1 | 2 | 7 | 27 | 19 | 16 | 8 | 7 |
| 10 | New Zealand | 27 | 18 | 4 | 10 | 19 | 13 | 11 | 6 | 1 |

*Source: 2009 Legatum Prosperity Index, http://www.prosperity.com/rankings.aspx*

## LIVES, LIFESTYLES, AND SOCIO-ECONOMIC SYSTEMS

Our lives and lifestyles are defined by habits, culture, conventions, prices, and income levels in the context of a socio-economic capitalistic system, which encourages production and consumption and is predicated on continuous growth.[45]

Moreover, a consumer-oriented culture, incessantly promoted through advertising and marketing, pervades all aspects of life in rich countries.[46] Powerful multinational corporations are the beneficiaries of this system and encourage further propagation of this culture globally. With powerful brands and massive marketing budgets, they do their best to influence not only our tastes and our choices, but also broadly encourage more consumption.[47] Money may not buy happiness, but in our materialistic culture, it comes close.

One might argue that multinational corporations even try to define our values and our needs.[48] How else can one explain the demand for nonessential luxury items, expensive branded products, mega-sized homes, yachts, and private planes? Why else would anyone market cigarettes, sugar-loaded fizzy drinks, or unhealthy fast food to consumers?[49]

Such discussions, while intellectually interesting, can sidetrack us from our main focus.[50] Returning to *energy services* that support the basic necessities of life, it is clear that culture and convention play critical roles. The British, to use a stereotype example, are accustomed to drinking their beer at room temperature while Americans like it ice cold. Americans like their showers hot while the British may put up with lukewarm water. Americans, on average, are used to keeping their homes warmer in the winter than the British.[51]

Conditioning is another important factor. Older generations that survived through the Great Depression and the rationing of the Second World War tend to be frugal, using less of everything even when they can afford it. The new generation grown during an age of plenty, rising standards, and relative affluence tend to be frivolous spenders and energy hogs by comparison.

Attitudes, however, continue to change, particularly among the educated and upwardly mobile, who exhibit increased awareness about the environment, are fond of organic food, recycling, and back-to-nature lifestyles. These powerful demographic and attitude shifts have not been lost on profit-seeking companies who are increasingly catering to these affluent consumers with a growing array of green and organic products.[52]

Social norms and income levels are other important determinants of energy use. For example, Americans, on average, live in much larger homes, own bigger cars, and drive longer distances while paying relatively lower gasoline prices.[53] Energy prices and their relation to average household disposable income play an important role in energy consumption patterns. Everything else being equal, people in Kentucky, West Virginia, or Wyoming—low-cost electricity states—can be expected to use significantly more electricity on a per capita basis than those in California or New York (Table 1). For the same reason, people in California and New York can be expected to be far more receptive to invest in energy efficiency measures than those in Kentucky, West Virginia, or Wyoming.

Similarly, price differentials for gasoline including taxes, explain why people in some countries own bigger and more cars and drive them farther—even accounting for population density and income levels.[54] If energy becomes scarce and more expensive—for example through the introduction of a carbon tax—energy consumption can be expected to drop, which is why many economists favor economy-wide, technology-neutral carbon taxes as a means of combating climate change.[55]

Difference in income levels is another factor that explains differences in energy consumption levels. For high-income earners, energy constitutes a relatively small fraction of the disposable income while the opposite is true for low-income consumers. If energy prices continue to rise, as some experts predict, the percentage of average disposable income devoted to energy expenses are likely to become more pronounced. If carbon taxes are introduced, either directly or indirectly, this will add to energy budgets for petrol, electricity, and heating. Sooner or later, consumers—and politicians—will notice.

OFGEM, the United Kingdom's energy regulator, for example, estimates that by 2016, the average annual electricity and household gas bill in Britain could rise by 60% to £2,000 (approx. $3,100), or roughly 10% of average household disposable income. How would consumers react to higher energy bills? Utility executives believe that they will have to get used to the idea. But will they? *Fuel poverty*, already a major political issue in Britain, is likely to become more pronounced if OFGEM's predictions materialize.

In his book *$20 Per Gallon*, Christopher Steiner (2009) envisions a future where petrol may cost the consumer $20 per gallon and examines how peoples' lives, lifestyles, and livelihoods may be affected. It is hard to imagine anything with a more dramatic impact on life and lifestyles than the price of energy—because it affects the price of everything else, from the cost of transportation to food to nearly all other products and services we need and use.

The book's most interesting contribution, however, is to point out the many *benefits* of expensive oil once we make the necessary adjustments. Steiner portrays a future where we *could* enjoy better lifestyles on a fraction of energy we currently use. One may not agree

with his vision or conclusions but it is an intriguing and promising perspective.

The point of his arguments ... is that lifestyles, habits, and norms are adjustable and can be *influenced* over time. Plastic water bottles, once fashionable, have become the curse of environmentalists. More shoppers bring reusable bags to carry groceries home. Some stores actually charge for them or give token rewards if you bring your own bags. More aluminum cans are recycled. As these examples illustrate, humans are adaptable to change, and given the means and the motivations, can be influenced to do the right things.

## TOWARD A MORE SUSTAINABLE LIFESTYLE

The examples and anecdotes offered thus far lead to three important observations:

- The amount of energy required to support basic human needs is *not* a preordained or set number.
- The amount of energy required to sustain a given lifestyle can be *adjusted* through substitution, investment, by adjusting prices, influencing habits, social and cultural norms, policies, and standards.
- Markets, while powerful and efficient, need occasional fine-tuning, prodding, and appropriate incentives—and such meddling can be supported so long as the end justifies the means.[56]

The main instruments of change, illustrated through examples, include:

- Investment and substitution—more capital and superior technology, all else being equal, typically results in lower energy usage with substantial savings over extended life of the energy using stock.
- Higher prices will result in reduced energy consumption and encourage substitution

Table 1    Average Retail Electricity Prices Among US States

### Want Cheap Juice?

AVERAGE RETAIL AND RESIDENTIAL RATES*, CENTS/kWH, 12 MONTH AVERAGE ENDING JUNE 2006

| REGION/STATE | AVG. RETAIL RATE | AVG. RESIDENTIAL RATE | REGION/STATE | AVG. RETAIL RATE | AVG. RESIDENTIAL RATE |
|---|---|---|---|---|---|
| New England | 13.29 | 14.93 | MD | 8.58 | 8.60 |
| CT | 13.20 | 15.02 | NC | 7.42 | 8.94 |
| ME | 10.63 | 13.98 | SC | 6.89 | 9.75 |
| MA | 14.04 | 15.44 | VA | 6.74 | 8.31 |
| NH | 13.65 | 14.44 | WV | 5.05 | 6.22 |
| RI | 13.32 | 14.49 | East South Central | 6.55 | 7.82 |
| VT | 11.22 | 13.27 | AL | 6.85 | 8.41 |
| Mid Atlantic | 11.20 | 13.00 | KY | 5.18 | 6.74 |
| NJ | 11.33 | 12.11 | MI | 8.27 | 9.44 |
| NY | 13.81 | 16.49 | TN | 6.68 | 7.35 |
| PA | 8.48 | 10.19 | West South Central | 9.18 | 10.82 |
| East North Central | 7.23 | 8.81 | AK | 6.54 | 8.29 |
| IL | 7.04 | 8.43 | LA | 8.61 | 9.38 |
| IN | 6.20 | 7.89 | OK | 7.36 | 8.47 |
| MI | 7.92 | 9.20 | TX | 9.96 | 11.88 |
| OH | 7.38 | 8.94 | Mountain | 7.40 | 8.84 |
| WI | 7.81 | 10.07 | AZ | 7.94 | 9.04 |
| West North Central | 6.51 | 7.83 | CO | 7.84 | 9.18 |
| IA | 6.88 | 9.51 | ID | 5.14 | 6.35 |
| KS | 6.79 | 8.11 | Mont | 6.81 | 8.20 |
| MN | 6.81 | 8.51 | NV | 9.26 | 10.57 |
| MO | 6.18 | 7.25 | NM | 7.64 | 9.25 |
| ND | 5.99 | 7.12 | UT | 6.02 | 7.63 |
| SD | 6.66 | 7.41 | WY | 5.16 | 7.52 |
| South Atlantic | 7.98 | 9.25 | Pacific | 9.93 | 10.72 |
| DL | 8.14 | 9.68 | CA | 11.89 | 13.00 |
| DC | 9.57 | 9.33 | OR | 6.38 | 7.37 |
| FL | 9.56 | 10.39 | WA | 5.90 | 6.65 |
| GA | 7.73 | 8.96 | Hawaii | 20.06 | 22.55 |
|  |  |  | US Avg. | 8.51 | 9.92 |

*These are averages for all applicable retail and residential rates, respectively, for each state including IOUs, munis and co-ops. Average rates are weighted by taking the total revenues and total kWh sales for each company. Non-IOU data is from EIA, No data is provided for Alaska, nor Nebraska, which has no IOUs

Source: Typical Bills and Average Rates, Oct 07, Edison Electric Institute

of other input variables such as labor and capital.[57] Fuel and carbon taxes,[58] for example, are among well-known means of discouraging energy use in general and carbon-heavy forms of energy, specifically. Moreover, rising price signals offer powerful incentives in directing investment and resources to address scarcities as they are encountered.

- Norms and habits are amenable to gradual change through education, conditioning, and with the emergence of superior alternatives and substitutes.[59] Recent popularity of CFLs, recycling aluminum cans, and reusable grocery bags are good examples.
- Energy policies can induce gradual change,[60] for example, by switching to low-carbon fuels and more efficient utilization of resources.
- Standards are among the most potent means of encouraging efficient utilization of energy and other scarce resources.
- There is increased recognition of the role of governments to control or modify private sector investments in ways that increases human welfare while reducing wasteful energy consumption.

Viewed in this context, the energy and sustainability problem, which appears insurmountable at first glance, may in fact be amenable to *adjustments* at *multiple* levels. Granted, the energy infrastructure is long-lasting and formed habits are hard to change, but persistent application of available instruments over time will deliver substantive results.

Take the case of zero net energy buildings, mandated in the state of California starting in 2020 for *new* homes and 2030 for *new* commercial buildings. Admittedly, the impact of the new regulation will be modest at first since it only applies to new construction, but will become

pronounced over time. The same applies to the 33% renewable mandate in California by 2020. Opponents of such measures complain about the extra costs and the fact that it will drive more business away from the state. Proponents believe that these requirements will eventually create more jobs than they destroy, will encourage technological innovation, and benefit the environment in the long run. California's climate bill, which requires state-level GHG emissions to be reduced to 1990 levels by 2020, have also come under attack by the same critics but are supported by the proponents on its positive merits.

While the problem of sustainability is not trivial and there are no simple single silver bullets, there is enormous *flexibility* in how we can meet our basic human needs, and in this sense, I prefer to view the glass as half full rather than half empty.

## NOTES

1. Another colorful quote attributed to Gandhi is "Live simply so others can simply live."
2. The current head of the Inter-governmental Panel on Climate Change (IPCC), Dr. Ravindra Pachauri, a vegetarian, has proposed that rich countries should consider going meatless for one day per week as a way to reduce human impact on the environment.
3. This, of course, applies to rich countries where a lot of energy is used on a per capita basis.
4. The wasted heat, of course, is appreciated in colder climates as a useful by-product of light.
5. As explained by Long et al. even today, private utilities in many parts of the world have perverse incentives to sell more, rather than less—which means they have little or no interest in encouraging energy conservation.

6. This is among a number of the market *failures* or *barriers* to energy efficiency often mentioned in the literature, including Long et al.

7. Meyer et al. illustrate a few examples of tax codes and prevailing laws that encourage more consumption, rather than conservation.

8. There are a number of persistent obstacles that prevent the optimal choice of investments in favor of lowest up-front costs as opposed to life-cycle costs, including energy costs over the long life of appliances, buildings, cars, and other devices.

9. This leads to philosophical and ideological arguments on whether consumers in fact need protection in the form of minimum energy efficiency standards and/or energy efficiency labeling or should we leave it to buyers to rely on their own wits and motivations.

10. Ehrhardt-Martinez et al., refers to this phenomenon.

11. Meyer et al. provide examples of tax codes and pricing tariffs that encourage consumption.

12. For a discussion of what is an appropriate price, refer to Sioshansi, F. (2010). What is the right price? In D. Reeves (Ed.), *Current Affairs—Perspectives on Electricity Policy for Ontario.* University of Toronto Press.

13. Estimate from U.S. Environmental Protection Agency (EPA) reported in *The Wall Street Journal,* "Encouraging Business to Turn Off the Lights," 27 April 2010.

14. Arent et al. describe the prospects for low-carbon renewable energy resources.

15. Bollino et al. refer to the significance of positional goods and the element of style and prestige associated with many personal consumption decisions. For example, if driving a private car is perceived as superior to riding a bike or taking public transit, this may influence the ultimate choice.

16. Some cities are experimenting with offering free public bicycles that people can take from one point to another and leave at designated racks for others to use. Others have devised inexpensive bike rental systems where users can pay by the hour using a credit or debit card.

17. Sovacool provides an example where city planners can essentially ration use of private cars in city centers by making parking scarce and expensive. Residents of big congested cities such as New York, London, or Tokyo are better off without a private car for lack of parking and congestion of inner-city traffic.

18. The city of Bogota in Colombia has specially designated lanes for express buses that whisk passengers around while traffic snarls force private cars to sit and wait in endless traffic jams.

19. A few cities, including London, Stockholm, and Singapore, have introduced access charges within the central business district (CBD) while others ban private vehicles from certain areas altogether.

20. Car-obsessed California has recently introduced measures to meet its ambitious climate change target that requires statewide emissions to be reduced to 1990 levels by 2020.

21. Moran provides counter-arguments in describing evolving traffic patterns in urban centers.

22. Of course, one must consider the energy and pollution to generate the electricity that is stored in car batteries and their associated conversion losses.

23. According to one estimate, the percentage may be as high as 20% in the U.S.

24. Of course, as usual, there is no free lunch. In this case, Internet service providers and data centers consume energy to maintain the infrastructure, but the net energy savings is still substantial relative to the alternatives.

25. Refer to Gray and Zarnikau and Prindle and Finlinson for further discussion of residential and commercial energy use.

26. California's three investor-owned utilities, with the support of state regulators, have sold over

95 million CFLs at subsidized prices to consumers, roughly three per capita.

27. A number of countries in Europe as well as Australia have banned the use of incandescent lights altogether, further increasing the demand for CFLs.

28. Refer to Brown et al. for a discussion of energy efficiency opportunities in the industrial sector.

29. Prindle and Finlinson address ways of overcoming these barriers.

30. Bauermann et al. discusses district heating.

31. Sovacool describes how the planners for the carbon-neutral city of Masdar in UAE are employing many design elements used in traditional buildings in hot, arid areas.

32. In practice, of course, one must strive to minimize overall resource use, not *just* energy.

33. Buijs describes the significance of rapidly developing countries such as China.

34. Refer to Trainer for a discussion of this.

35. Moran compares and contrasts these differences.

36. For an evolutionary perspective on how various societies have evolved through history, refer to Jared Diamond's *Guns, Germs and Steel* (1999).

37. Refer to Bartiaux et al. (Chapter 3) for an anthropological perspective on the concept of needs, wants, and desires.

38. Refer to Chapter 16 by Stulz et al. for a discussion of energy use in Switzerland and efforts to reduce it.

39. According to 2009 Legatum Prosperity Index, Switzerland is ranked as the second "most prosperous" country; the U.S. is ranked #9.

40. Gray and Zarnikau, discuss the cost implications of moving to zero net energy buildings.

41. Ehrhardt et al., covers the so-called "beyond the meter" applications that enable reduced energy consumption through improved utilization and management.

42. Clymer, describes some of the measures pursued in the case of city of Austin, Texas.

43. Gray and Zarnikau, and Rajkovich et al., describe the concept of zero net energy.

44. Energy efficiency guru Amory Lovins lives in a zero net energy home using the most advanced appliances and devices on the "bleeding edge" of technology. But these concepts are still beyond the means and reach of average homeowners.

45. A number of economists are questioning the necessity of continuous economic growth including Tim Jackson in his book *Prosperity Without Growth: Economics for a Finite Planet* (Earthscan, 2009) and Peter Victor in his book *Managing Without Growth: Slower by Design, Not Disaster* (Edward Elgar, 2008).

46. Trainer offers further discussion of the evils of what he considers excessive consumerism.

47. But as Trainer explains our capitalistic system not only affords the rich to live extravagant lifestyles but encourages more of us to aspire to reach similar status.

48. This, of course, is a circular argument, since profit-seeking companies can only sell what consumers will willingly buy.

49. Moran questions consumer "paternalism" pointing out that consumers are smarter than many consumer advocates give them credit for.

50. Another prominent feature of our current system, disturbing to many, is the vast money spent on arms, defense, and wars. World military spending doubled during the past decade, reaching $1.53 trillion in 2009, according to a Swedish study reported in *The Wall Street Journal*, 2 June 2010.

51. Refer to Gray and Zarnikau.

52. There are numerous examples of this trend, including General Electric's Ecomagination campaign, Wal-Mart's efforts to cater to green consumers, and Intel buying all its electricity from renewable resources.

53. Gray and Zarnikau provide comparisons.

54. Refer to Bollino and Polinori on the effect of energy taxes on energy consumption.

55. It must, however, be noted that elasticity of demand for energy tends to be low, meaning that significant increases in prices are required to produce rather small reductions in energy use.

56. Needless to say, there are considerable differences in views on how best to regulate markets and to what extent. For example, refer to Meyer et al., who prefer a strong hand for government in controlling the excesses of free markets.

57. The elasticity of demand for energy appears to be low, however, suggesting that relatively large price increases may be necessary to promote energy efficiency or substitution. This suggests that higher prices, combined with other incentives, may be needed to affect consumption and behavior.

58. Refer to Bollino and Polinori, for more on petrol taxes.

59. Prindle and Finlinson, for example, examine how college students living in dormitories may be persuaded to take 3-minute showers instead of the current 14-minute variety.

60. Felder et al., describes alternative energy futures.

# THE CLIMATE CRISIS AT THE END OF YOUR FORK

By Anna Lappé

## TAKING A BITE OUT OF CLIMATE CHANGE

On September 8, 2008 Dr. Rajendra Pachauri, chair of the United Nations Intergovernmental Panel on Climate Change, spoke to 400 people gathered for an event hosted by the animal welfare organization Compassion in World Farming.[1] Pachauri, an Indian economist (and vegetarian) who had just been reelected to a second term as chairman, made one of the most public and bold statements about the connection between our diet and global warming on the world stage: Choosing to eat less meat, Pachauri said, or cutting out meat entirely, is one of the most important personal choices we can make to address climate change.

"In terms of immediacy of action and the feasibility of bringing about reductions in a short period of time, it clearly is the most attractive opportunity," said Pachauri. "Give up meat for one day [a week] initially, and decrease it from there."

To many of us, Pachauri's specific prescription for addressing change might come as a surprise. When we think about the culprits behind the climate crisis, we tend to think about Big Oil or dirty coal-fired powered plants. We picture cars and industrial skylines, or imagine factories and smokestacks. It's time we start thinking about another sector of the economy that is increasingly exacerbating the climate crisis. The global food system—including deforestation to make way for crops for cattle and cars—is responsible for an estimated one-third of total greenhouse gas emissions (Table 1: Main Sources of Emission from Agriculture).

Livestock production alone contributes to 18 percent of the global warming effect—more than the emissions from every single car, train, and plane on the planet.[2] Though livestock production only contributes 9 percent of carbon dioxide emissions, the sector is responsible

Table 1    Main Sources of Emission from Agriculture

| MAIN SOURCES OF EMISSIONS | PERCENT OF THE *TOTAL* GLOBAL WARMING EFFECT OF EMISSIONS |
|---|---|
| **On the Farm** | |
| Fertilizer production and distribution | 1.5 to 2% |
| Methane and nitrous oxide emissions | 12% |
| **On the Land** | |
| Deforestation and other land use changes | 18% |
| **On the Road** | |
| Transportation emissions from seed to plate | *Specific food-system data unavailable* |
| **Additional Sources** | *Specific food-system data unavailable* |
| Waste and manufacturing | |
| **Estimated Total** | 33% of the total global warming effect can be attributed to the food system. |

*Note:* These percentages are a rough estimation of the global warming effect of the food system. Because of the complexity of global warming science as well as the challenge of securing accurate estimations of emissions from these various sectors, this estimation is the best we can hope for. While more research is certainly needed, in the meantime, we must begin to make policy changes to address what we do know: That our food system is a significant contributor to the global warming effect, and that we each can do something about it.

for 37 percent of methane and 65 percent of nitrous oxide, both potent greenhouse gases.[3]

Move over Hummer, say hello to the hamburger.

## THE FOOD AND CLIMATE CHANGE CONNECTION

So how is food—supposedly life-*sustaining* stuff—one of the key factors in an environmental crisis that threatens the basis of life on earth? A big part of the answer is in the rapid and radical twentieth- century transformation of our food system from sustainably based, locally focused production, to a fossil-fuel addicted industrialized system. Agriculture has changed more in than the past two generations than it did in the previous 12,000 years. Unfortunately for us, almost every single aspect of our modern industrial system creates greenhouse gas emissions. And, as Dr. Pachauri says, another big reason is the rapid growth of

livestock production. Indeed, to produce 2.2 pounds of beef burns enough energy to light a 100 watt bulb for twenty days Pachauri noted in his remarks earlier this month.[4]

## On the Farm

*Fertilizers and On-Farm Fossil Fuel Use* Industrial farms are fossil-fuel addicted places, from their reliance on fossil fuels for powering machinery to petroleum-based chemicals used to create artificial soil fertility, protect against pests, and stave off weeds. The use of fossil fuels on farms, as well as in the manufacture of fertilizers and agricultural chemicals, contributes to the emissions of the food sector

With one-third of the world's cereal harvest and 90 percent of the world's soy harvest being raised for animal feed, the energy required to grow those crops is a major factor in these on-farm emissions.[5] In the United States and Canada, half of all synthetic fertilizer is used for feed crops.[7] In the UK, the total is nearly 70 percent.[7].

A major reason that beef, in particular, has such a large environmental impact is because, of all livestock, cattle are among the worst converters of grain to meat. Whereas in nature cattle, which are ruminants, convert inedible-to-humans grasses into high- grade proteins, under industrial production grainfed cattle only provide about 1 pound of beef for every 10 to 16 pounds of feed they consume.[8]

Because industrialized agriculture also relies on huge amounts of water for irrigation, these farms will be much more vulnerable as climate change increases extreme droughts. Globally, 70 percent of the world's available freshwater is being diverted to irrigation- intensive agriculture.[9]

**The Livestock Liability** The other reason that livestock production has such a big impact on climate change is because livestock are among the main sources of the world's methane emissions. (Rice cultivation is another. According to the Stern Report, rice cultivation emits one-tenth of agricultural emissions.[10]) Ruminant livestock, including cattle, buffalo, sheep, and goats, are the main agricultural sources of methane. They can't help it; it's in their nature. Ruminants digest through microbial, or "enteric" fermentation, which produces methane that is then released by the animals through belching and, to a lesser degree via their, er, tailpipes. While this process enables ruminants to digest fibrous grasses that we humans can't convert into digestible form, it also contributes to livestock's climate change toll. Enteric fermentation accounts for one-quarter of the total emissions from the livestock sector. (Land use changes, such as deforestation and desertification caused by over-pasturing livestock or growing feed crops, account for another 35.4%, while manure accounts for 30.5%).[11]

There's another problem with industrial livestock: what happens with the waste. Now, in sustainable systems tapping into nature's wisdom, there is no such thing as waste: manure is part of a holistic cycle; it's fertilizer. But in confined animal feeding operations (CAFOs), waste is not cycled through the farm, there's too much of it. Instead, waste is stored in manure "lagoons," as they're euphemistically called. Without sufficient oxygenation, this waste ends up emitting methane and nitrous oxide. The United States scores at the top of the world for methane emissions from manure, and pigs are at top in terms of methane emissions, responsible for half of the globe's total.[12]

The sheer number of animals being raised for meat on the planet is another reason livestock production accounts for nearly one-fifth of all the globe's greenhouse gas emissions. In 1965, 10 billion livestock animals were slaughtered each year; today that number is 55 billion.[13]

## On the Land

The bulk of the pressure on land around the globe, from precious wetlands in Indonesia to rainforests in Brazil, comes from the agricultural sector.[14] These rainforests and wetlands play a vital role in climate stability because they sequester carbon, absorbing and storing carbon dioxide from the atmosphere in the soils and plants. With their destruction, carbon is released back into the atmosphere and the carbon cycle that keeps our climate in balance comes further unhinged. Because these lands play such a vital role as carbon sinks, it's no surprise that their destruction is partly responsible for the emissions from land use changes that add up to nearly 18 percent of the total global warming effect.

The biggest driver behind these "land use changes," as the climate change folks call them, is the expansion of pasture for cattle, feed crops for livestock, and oil palm production for processed foods and biofuels.[15] Most of these land use changes are concentrated in just a handful of countries. Brazil is the heart of rainforest destruction, mainly to meet demand for livestock grazing and feed. Malaysia and Indonesia are the world's main producers of oil palm, where plantations are leading to rapid rainforest and wetland habitat destruction. Malaysia produced 43 percent and Indonesia 44 percent of the world's total palm oil last year.[16] While data on the exact amount of land that has been converted into oil palm plantations is hard to come by, we know that demand for palm oil has soared in the last two decades, especially because of growing demand for edible vegetable oils from the world's top two importing countries, India and China, according to an assessment by the USDA's Foreign Agricultural Service.[17]

The erosion and deterioration of soils on industrial farms also releases greenhouse gases into the atmosphere. By destroying the natural soil fertility, and disturbing the soil through tillage, industrial farming also adds to the deterioration of soil and its carbon emissions.[18]

## On the Road: The Food Miles Question

And what about the distance food travels to get to our plates? Despite all the attention to food miles, emissions from food transport are not the biggest culprit behind the sector's impact on climate change.

While total emissions from transportation contribute to 13.1 percent of the global warming effect, this includes emissions from toting around all kinds of things—from people to pork chops.[19] Transport emissions specifically from food are just a sliver of these emissions.

So why do food miles matter? Even though transportation isn't the main source of our morsels' emissions, reducing our food miles does make a dent in our dinner's emissions toll. Consider the fossil fuels wasted carting fresh tomatoes to New Jersey, a state with ample farmland that exports tons of tomatoes every year. Researchers at Rutgers University estimated that meeting the New Jersey demand for just one year's supply of out-of-state tomatoes for the state used up enough fossil fuel to drive an 18-wheeler around the world 249 times.[20]

Food miles matter because so much of our food transport is unnecessary. Local food is also a better choice because it is fresher and therefore healthier for our bodies. Buying local also supports our local economies. Purchasing local foods means supporting small-scale businesses and protecting green space in our communities.

Of course, direct trade with small-scale farmers in far away places can be a critical way to support economic development half-way around the world. But most of the global food trade isn't benefiting small-scale farmers, it's benefiting the biggest grain traders at the cost of the climate.[21] And most of this global food trade is completely unnecessary, or redundant. Consider, for example, fish caught off the coast of Maine, flash frozen, shipped to China for processing into filets, and shipped back to our mega-markets in the United States. Or, consider the business of beef. In 2007, the United States exported one 1.431 billion pounds of beef and veal (5.4 percent of our total beef production)[22] and imported 3.052 billion pounds of the same, measured by commercial carcass weight.[23] This cross-continent transport of food makes economic sense only because the true costs of such transport, including the big bill for its contribution to climate change, are not counted on the balance sheets of food corporations.

The main reason for sticking with the locavores is that the local food being celebrated is often a pseudonym for sustainably raised foods—and those foods will be the ones produced without the fertilizers, grown without destroying precious wetlands or rainforests, and with animals raised on pasture, not in confinement.

*Other Sources of Emissions* Because the food system is connected to so many aspects of our lives, it's hard to get an exact accounting of total emissions. Here are some other places where the food system's emissions crop up.

- **Waste**: Another 4 percent of global greenhouse gas emissions comes from waste, including food waste. Where does all our uneaten food end up and the tons of food ready for harvest that never even makes it to our plates? Landfills. And land-

fills are a key source of methane, as food and other refuse decay.

- **Manufacturing**: A further 10 percent of global emissions stems from manufacturing and construction, which includes construction for the food industry.

## HOW CAN FARMING HELP ADDRESS THE CLIMATE CRISIS?

In nature, plants transform the sun's energy into food that provides a foundation for life. We humans are fueled by this transformation either directly (we eat the food) or indirectly (we eat the animals that have fed on this energy). It's a clever cycle: it's inherently abundant. But the industrialization of agriculture, picking up pace

Table 2   An Overview of Food System Sources of Greenhouse Gases

| Gas | What's Food Got to Do With It? A summary of causes | Global Warming Potential Relative to Carbon Dioxide |
| --- | --- | --- |
| Carbon Dioxide | Carbon dioxide is emitted by burning fossil fuels to power farm machinery, produce agricultural chemicals and transport food. In addition, carbon dioxide is released when forests and wetlands are cleared for crop production, particularly for animal feed, pasture, or oil palm production. Finally, trapped carbon in soils is also released through soil erosion and deterioration on industrial farms. | |
| Methane | Agricultural methane is released primarily by ruminants, such as cattle, and during rice production. | 23 times the greenhouse effect of carbon dioxide over 100 years. 62 times over 20 years.[†] |
| Nitrous Oxide | Nitrous oxide is released mainly from the use of man-made fertilizer, especially the overuse of nitrogen on crops grown for animal feed. | 296 times the greenhouse effect of carbon dioxide over 100 years. 275 over 20 years.[‡] |

*A note about methane and nitrous oxide*: In the U.S., the food system accounts for an estimated 17 percent of all fossil fuel use, but accounts for a much larger percentage of our country's methane and nitrous oxide emissions. Widespread overuse of artificial nitrogen fertilizer, for instance, much of which is wasted in leaching and runoff, contributes to three-quarters of our country's nitrous oxide emissions. And globally, agriculture is responsible for nearly two-thirds of methane emissions. While methane and nitrous oxide make up much smaller portions of total greenhouse gas emissions, these gases are still important factors in the climate crisis, in part because they each have stronger global warming effects than carbon dioxide and also because they are an increasing portion of total emissions.

[†] Steinfeld, H., P. Gerber, et al. (2006). *Livestock's Long Shadow: Environmental Issues and Options*. Rome, Food and Agriculture Organization of the United Nations.

[‡] Ibid.

Table 3   The Core Differences Between Climate-Crisis Agriculture and Climate-Friendly Farming

| CLIMATE-CRISIS AGRICULTURE | CLIMATE-FRIENDLY FARMING |
|---|---|
| Input-Intensive | Knowledge-Intensive |
| Ignores Place | Values Place |
| Emits Carbon | Stores Carbon |
| Dependant on Chemicals | Depends on Nature for Fertility and Pest Management |
| Disrupts Natural Cycles | Protects Nature's Cycles |
| Squanders Energy | Produces Energy |
| Fossil-Fuel Powered | People and Animal Powered |
| Wastes Water | Retains and Conserves Water |

in the past generation, has flipped the natural abundance of farming on its head. Instead of producing energy, industrial agriculture consumes it, through the addiction to fossil fuel-powered machinery and petroleum-based agrochemicals. Industrial farms are often considered highly efficient, but only because these wasted inputs and devastating outputs—including the impact on climate change—are not accounted for. (See Table 3: The Core Differences between Climate-Crisis Agriculture and Climate-Friendly Farming).

Unlike industrial farms, small-scale organic and sustainable farms rely on people power, not heavy machinery, and depend on nature, not manmade chemicals for soil fertility and to handle pests. As a result, small-scale sustainable farms have been found to emit between one-half and two-thirds less carbon dioxide for every acre of production.[24]

New research is documenting that organic farms can emit as much as half the carbon dioxide as chemical farms. Organic farms also use much less fossil fuel energy than their conventional counterparts, in many cases as much as one-third less, and studies are also showing that organic farming can sequester carbon, providing a potentially powerful tool to help us address climate change. In fact, 10,000 medium-sized organic farms can store as much carbon in the soil as we would save if we took one million cars off the road.[25]

Yes, the very source of energy—the food consumers eat—has become one of the planet's worst contributors to climate instability, but it need not be. There is another way.

## WHAT YOU CAN DO

### Eat and Drink with the Environment in Mind

Many of the resources at Sustainable Table can help you make climate-friendly food choices every day. Here are some of the principles of low-carbon dining, with links to more information.

- Eat less meat and dairy: Go cold turkey, or just trim your consumption. Try cutting out meat just one day of the week. Check out Meatless Monday at www.meatless-monday.com for more inspiration.
- **Choose organic and sustainably raised foods:** Visit the Eat Well Guide at www.eatwellguide.org to find local food near your home.
- **Eat local:** Lower your food mile odometer, and always try to look for local sustainable food. Get inspired with these ideas from Sustainable Table (www.sus-

tainabletable.org/issues/buylocal) and by the original locavores (www.locavores.com).

- **Eat whole foods:** Cut back on processed foods and check out your "carbon foodprint" at Bon Appetit's Low-Carbon Diet Calculator at www.eatlowcarbon.org.

- **Take back the tap:** Kick the bottled water habit. Visit Food & Water Watch at www.fwwatch.org/water/bottled for more information about the environmental impact of bottled water, the benefits of tap water, and how you can add your voice to those of people across the country demanding better regulation of our precious public resource—water.

## Get Involved on a Broader Scale

Go beyond your plate to get involved with exciting campaigns that are making the food and climate connection. Here are three recommendations.

- **Rainforest Action Network's Agribusiness Campaign:**
  Join Rainforest Action Network's (www.ran.org) agribusiness campaign and learn what you can do to help stop the deforestation of the world's precious rainforests.

- **Center for Food Safety's Cool Foods Campaign:**
  Check out the Center's newest campaign, helping eaters and businesses around the country learn about the simple choices they can make to support a climate-friendly food system (www.coolfoodscampaign.org).

- **La Via Campesinas's Small Farmers Cooling the Planet:**
  Join with this international network of farmer associations across the planet to promote small-scale farming as a key

strategy for mitigating and adapting to climate change (www.laviacampesina.org).

## Get Informed

- Visit Take a Bite out of Climate Change at www.takeabite.cc/learn to get more resources and sign up to stay informed.

- Learn more about the meat and climate change connection. Read the definitive study from the United Nations: Livestock's Long Shadow (www.fao.org/docrep/010/a0701e/a0701e00.htm).

- Check out the Manifesto on Food and Climate Change at www.future-food.org from Vandana Shiva and her colleagues at the International Commission on the Future of Food. The manifesto summarizes the connections between industrial agriculture and climate change and offers a compelling call to action from the perspective of farmers in the global south.

- Read Soil not Oil (www.southendpress.org/2008/items/87828) by Vandana Shiva from South End Press: Shiva connects the dots between the world's most pressing crises—food insecurity, oil dependence and climate change.

## REFERENCES

Eldredge, N. 2002. *Life on Earth: An Encyclopedia of Biodiversity, Ecology, and Evolution.* Santa Barbara, Calif., ABC-CLIO.

IPCC. 2007. *Climate Change 2007: Fourth Assessment Report of the Intergovernmental Panel on Climate Change.* New York, Cambridge University Press.

Lappé, F. M. 1991. *Diet for a Small Planet.* New York: Ballantine Books.

McMichael, J., J. Powles, et al. 2007. "Food, Livestock Production, Energy, Climate Change, and Health." The *Lancet* 370: 1253–1263.

Roberts, P. 2008. *The End of Food. Boston*: Houghton Mifflin Company.

Steinfeld, H., P. Gerber, et al. 2006. *Livestock's Long Shadow: Environmental Issues and Options*. Rome, Food and Agriculture Organization of the United Nations.

Stern, N. H. and Great Britain Treasury. 2007. *The Economics of Climate Change: The Stern Review*. New York: Cambridge University Press.

Weis, A. 2007) *The Global Food Economy*: The Battle for the Future of Farming. London: Zed Books.

## SOURCES

1. "Global Warning—The impact of meat production and consumption on climate change," Dr Rajendra Pachauri, Chair of the Intergovernmental Panel on Climate Change. Speech at Compassion in World Farming, London, September 8, 2008.

2. Steinfeld, H., P. Gerber, et al. (2006). *Livestock's Long Shadow: Environmental Issues and Options*. Rome, Food and Agriculture Organization of the United Nations. Transportation is responsible for a total of 13 percent of the global warming effect.

3. Ibid. p. 79. See, for instance, http://www.fao.org/ag/magazine/0612sp1.htm.

4. "Global Warning—The impact of meat production and consumption on climate change," Dr Rajendra Pachauri, Chair of the Intergovernmental Panel on Climate Change. Speech at Compassion in World Farming, London, September 8, 2008.

5. "Global Warning—The impact of meat production and consumption on climate change," Dr Rajendra Pachauri, Chair of the Intergovernmental Panel on Climate Change. Speech at Compassion in World Farming, London, September 8, 2008.

6. Steinfeld, H., P. Gerber, et al. (2006). Livestock's Long Shadow: Environmental Issues and Options. Rome, Food and Agriculture Organization of the United Nations.

7. Ibid.

8. "Conversion ratios" are hotly contested. In Frances Moore Lappe's (Anna's mother's) book, *Diet for a Small Planet* (*Ballantine*, she estimated the conversion ratio for U.S.-feedlot cattle is sixteen pounds of grain and soy to produce one pound of beef. In *The End of Food*, journalist Paul Roberts argues that these conversion ratios don't account for the 60 percent of a cow's weight that is bone, organ, and hide—inedible stuff. The real conversion ratio for beef, Roberts argues should be even lower. He estimates that it takes "a full 20 pounds of grain to make a single pound of beef." From Roberts, P. (2008). The End of Food. Boston, Houghton Mifflin Company. p. 293. See also Lappé, F. M. (1991). *Diet for a Small Planet*. New York, Ballantine Books.

9. See for instance, Eldredge, N. (2002). *Life on Earth: An Encyclopedia of Biodiversity, Ecology, and Evolution*. Santa Barbara, Calif., ABCCLIO. Viewable at http://www.landinstitute.org/vnews/display.v/ ART/2002/08/23/439bd36c9acf1.

10. Stern, N. H. and Great Britain. Treasury. (2007). *The Economics of Climate Change: The Stern Review*. Cambridge, UK ; New York, Cambridge University Press. See Annex 7g. (www.hm-treasury.gov.uk/independent_ reviews/stern_review_economics_climate_change/sternreview_index.cfm)

11 McMichael, J., J. Powles, et al. (2007). "Food, Livestock Production, Energy, Climate Change, and Health." The Lancet 370: 1253–1263.

12. Steinfeld, H., P. Gerber, et al. (2006). *Livestock's Long Shadow: Environmental Issues and Options*. Rome, Food and Agriculture Organization of the United Nations.

13. United Nations FAO, quoted in: Weis, A. (2007). The Global Food Economy: The Battle for the Future of Farming. London, Zed Books. p. 19.

14. Stern, N. H. and Great Britain. Treasury. (2007). *The Economics of Climate Change: The Stern Review*. Cambridge, UK ; New York, Cambridge University Press. 539.

15. Ibid. p. 539.

16. USDA FAS, "Indonesia: Palm Oil Production Prospects Continue to Grow," December 31, 2007. Viewable at www.pecad.fas.usda.gov/highlights/2007/12/Indonesia_palmoil/.

17. USDA FAS, "Indonesia: Palm Oil Production Prospects Continue to Grow," December 31, 2007. Total area for Indonesia palm oil in 2006 is estimated at 6.07 million hectares according to a information from the Indonesia Palm Oil Board (IPOB). Viewable at http://www.pecad.fas.usda.gov/highlights/2007/12/Indonesia_palmoil/.

18. Stern, N. H. and Great Britain. Treasury. (2007). *The Economics of Climate Change: The Stern Review*. Cambridge, UK ; New York, Cambridge University Press.

19. IPCC (2007). *Climate Change 2007: Fourth Assessment Report of the Intergovernmental Panel on Climate Change*. New York, Cambridge University Press. See Figure 1, Chapter 2.

20. From a Rutgers University study that found that 635,000 gallons of fuel was needed annually to import tomatoes into New Jersey, generating 6,616 metric tons of carbon dioxide, requiring nearly a 1.5 square miles of forest to absorb.

21. See for example, the USDA ERS, "Processed Food Trade Pressured by Evolving Global Supply Chains," www.ers.usda.gov/AmberWaves/February05/Features/ProcessedFood.htm.

22. Data from USDA ERS, U.S. Cattle and Beef Industry, 2002–2007, http://www.ers.usda.gov/news/BSECoverage.htm

23. U.S. Red Meat and Poultry Forecasts. Source: World Agricultural Supply and Demand Estimates and Supporting Materials. From USDA ERS. For further information, contact: Mildred Haley, (202) 694–5176, mhaley@ers.usda.gov. See also www.ers.usda.gov/Browse/TradeInternationalMarkets/.

24. IPCC (2007). *Climate Change 2007: Fourth Assessment Report of the Intergovernmental Panel on Climate Change*. New York, Cambridge University Press.

25. See for instance studies from the Rodale Institute, found here: www.newfarm.org/depts/NFfield_trials/1003/carbonsequest.shtml.

# THE PLEASURES OF EATING

## WHAT CITY PEOPLE CAN DO

By Wendell Berry

**M**any times, after I have finished a lecture on the decline of American farming and rural life, someone in the audience has asked, "What can city people do?"

"Eat responsibly," I have usually answered. I have tried to explain what I mean by that, but afterwards I have invariably felt there was more to be said. I would like to attempt a better explanation.

I begin with the proposition that eating is an agricultural act. Eating ends the annual drama of the food economy that begins with planting and birth. Most eaters, however, are no longer aware that this is true. They think of food as an agricultural product, perhaps, but they do not think of themselves as participants in agriculture. They think of themselves as "consumers." If they think beyond that, they recognize that they are passive consumers. They buy what they want—or what they have been persuaded to want—within the limits of what they can get.

They pay, mostly without protest, what they are charged. And they mostly ignore certain critical questions about the quality and the cost of what they are sold: How fresh is it? How pure or clean is it, how free of dangerous chemicals? How far was it transported, and what did transportation add to the cost? How much did manufacturing or packaging or advertising add to the cost? When the food product has been manufactured or "processed" or "precooked," how has that affected its quality or price or nutritional value?

Most urban shoppers would tell you that food is produced on farms. But most do not know what farms, or what kinds of farms, or where the farms are, or what knowledge or skills are involved in farming. They apparently have little doubt that farms will continue to produce, but they do not know how or over what obstacles. For them, food is pretty much an abstract idea—something they do not know

157

or imagine—until it appears on the grocery shelf or on the table.

The specialization of production induces specialization of consumption. Patrons of the food industry have tended more and more to be mere consumers—passive, uncritical, and dependent. Indeed, this may be one of the chief goals of industrial production. The food industrialists have persuaded millions of consumers to prefer food that is already prepared. They will grow, deliver, and cook your food for you and (just like your mother) beg you to eat it. That they do not yet offer to insert it, prechewed, into our mouth is only because they have found no profitable way to do so. We may rest assured that they would be glad to find such a way. The ideal industrial food consumer would be strapped to a table with a tube running from the food factory directly into his or her stomach.

Perhaps I exaggerate, but not by much. The industrial eater no longer knows or imagines the connections between eating and the land, and is therefore passive and uncritical—in short, a victim. When food, in the minds of eaters, is no longer associated with farming and with the land, the eaters suffer a kind of cultural amnesia that is misleading and dangerous.

Like industrial sex, industrial eating has become a degraded, poor, and paltry thing. Our kitchens and other eating places more and more resemble filling stations, as our homes more and more resemble motels. "*Life is not very interesting,*" we seem to have decided. "*Let its satisfactions be minimal, perfunctory, and fast.*" We hurry through our meals to go to work and hurry through our work in order to "recreate" ourselves in the evenings and on weekends. And all this is carried out in a remarkable obliviousness to the causes and effects, the possibilities and the purposes, of the life of the body in this world.

One will find this obliviousness represented in virgin purity in the advertisements of the food industry, in which food wears as much makeup as the actors. If one gained one's whole knowledge of food from these advertisements (as some presumably do), one would not know that the various edibles were ever living creatures, or that they all come from the soil, or that they were produced by work. The passive American consumer, sitting down to a meal of pre-prepared food, confronts inert, anonymous substances that have been processed, dyed, breaded, sauced, gravied, ground, pulped, strained, blended, prettified, and sanitized beyond resemblance to any part of any creature that ever lived. The products of nature and agriculture have been made, to all appearances, the products of industry. Both eater and eaten are thus in exile from biological reality. And the result is a kind of solitude, unprecedented in human experience, in which the eater may think of eating as, first, a purely commercial transaction between him and a supplier and then as a purely appetitive transaction between him and his food.

And this peculiar specialization of the act of eating is, again, of obvious benefit to the food industry, which has good reasons to obscure the connection between food and farming. It would not do for the consumer to know that the hamburger she is eating came from a steer who spent much of his life standing deep in his own excrement in a feedlot, helping to pollute the local streams, or that the calf that yielded the veal cutlet on her plate spent its life in a box in which it did not have room to turn around. And, though her sympathy for the slaw might be less tender, she should not be encouraged to meditate on the hygienic and biological implications of mile-square fields of cabbage, for vegetables grown in huge monocultures are dependent on toxic chemicals—just as animals in close confinements are dependent on antibiotics and other drugs.

The consumer, that is to say, must be kept from discovering that, in the food industry—as in any other industry—the overriding concerns are not quality and health, but volume and price. For decades the entire industrial food economy has been obsessed with volume. It has relentlessly increased scale in order (probably) to reduce costs. But as scale increases, diversity declines; so does health; and dependence on drugs and chemicals increases. Capital replaces labor by substituting machines, drugs, and chemicals for human workers and for the natural health and fertility of the soil. The food is produced by any means or any shortcuts that will increase profits. And the business of the cosmeticians of advertising is to persuade the consumer that food so produced is good, tasty, healthful, and a guarantee of marital fidelity and long life.

It is possible, then, to be liberated from the husbandry and wifery of the old household food economy, but only by entering a trap (unless one sees ignorance and helplessness as the signs of privilege, as many apparently do). How does one escape this trap? Only voluntarily, the same way one went in: by restoring one's consciousness of what is involved in eating; by reclaiming responsibility for one's own part in the food economy. One might begin with the illuminating principle of Sir Albert Howard's *The Soil and Health*, that we should understand "the whole problem of health in soil, plant, animal, and man as one great subject." Eaters, that is, must understand that eating takes place inescapably in the world, that it is inescapably an agricultural act, and how we eat determines, to a considerable extent, how the world is used. This is a simple way of describing a relationship that is inexpressibly complex. To eat responsibly is to understand and enact, so far as we can, this

complex relationship. What can one do? Here is a list, probably not definitive:

1. Participate in food production to the extent that you can. If you have a yard or even just a porch box or a pot in a sunny window, grow something to eat in it. Make a little compost of your kitchen scraps and use it for fertilizer. Only by growing some food for yourself can you become acquainted with the beautiful energy cycle that revolves from soil to seed to flower to fruit to food to offal to decay, and around again. You will be fully responsible for any food that you grow for yourself, and you will know all about it. You will appreciate it fully, having known it all its life.

2. Prepare your own food. This means reviving in your own mind and life the arts of kitchen and household. This should enable you to eat more cheaply, and will give you a measure of "quality control."

3. Learn the origins of the food you buy, and buy the food that is produced closest to your home. The idea that every locality should be, as much as possible, the source of its own food makes several kinds of sense. The locally produced food supply is the most secure, freshest, and the easiest for local consumers to know about and to influence.

4. Whenever possible, deal directly with a local farmer, gardener, or orchardist. All the reasons listed for the previous suggestion apply here. In addition, by such dealing you eliminate the whole pack of merchants, transporters, processors, packagers, and advertisers who thrive at the expense of both producers and consumers.

5. Learn, in self-defense, as much as you can of the economy and technology of industrial food production. What is added to the

food that is not food, and what do you pay for those additions?

6. Learn what is involved in the best farming and gardening.

7. Learn as much as you can, by direct observation and experience if possible, of the life histories of the food species.

The last suggestion seems particularly important to me. Many people are now as much estranged from the lives of domestic plants and animals (except for flowers and dogs and cats) as they are from the lives of the wild ones. This is regrettable, for there is such pleasure in knowing them. And farming, animal husbandry, horticulture, and gardening, at their best, are complex and comely arts; there is much pleasure in knowing them, too.

It follows that there is great displeasure in knowing about a food economy that degrades and abuses those arts and those plants and animals and the soil from which they come. For anyone who does know something of the modern history of food, eating away from home can be a chore. My own inclination is to eat seafood instead of red meat or poultry when I am traveling. Though I am by no means a vegetarian, I dislike the thought that some animal has been made miserable in order to feed me. If I am going to eat meat, I want it to be from an animal that has lived a pleasant, uncrowded life outdoors, on bountiful pasture, with good water nearby and trees for shade. And I am getting almost as fussy about food plants. I like to eat

vegetables and fruits that I know have lived happily and healthily in good soil, not the products of the huge, bechemicaled factory-fields that I have seen, for example, in the Central Valley of California. The industrial farm is said to have been patterned on the factory production line. In practice, it looks more like a concentration camp.

The pleasure of eating should be an extensive pleasure, not that of the mere gourmet. People who know the garden in which their vegetables have grown and know that the garden is healthy and remember the beauty of the growing plants, perhaps in the dewy first light of morning when gardens are at their best. Such a memory is one of the pleasures of eating. The knowledge of the good health of the garden relieves and frees and comforts the eater. The same goes for eating meat. The thought of the good pasture and of the calf contentedly grazing flavors the steak. Some, I know, will think of it as bloodthirsty or worse to eat a fellow creature you have known all its life. On the contrary, I think it means that you eat with understanding and with gratitude. A significant part of the pleasure of eating is in one's accurate consciousness of the lives and the world from which food comes. The pleasure of eating, then, may be the best available standard of our health. And this pleasure, I think, is pretty fully available to the urban consumer who will make the necessary effort.

# HUMAN HEALTH AND WELL-BEING IN AN ERA OF ENERGY SCARCITY AND CLIMATE CHANGE

By Cindy L. Parker and Brian S. Schwartz

In the past hundred years, we have created lifestyles, communities, food systems, water systems, transportation systems, and health systems that are entirely reliant on cheap and plentiful oil and that assume a favorable and stable climate. Our health and well-being have been shaped by these lifestyles and systems, but they have not necessarily been well served: Climate change and the threat of energy scarcity now pose serious challenges to our "health system," specifically health care services and public health services.

The consequences of climate change and energy scarcity will be wide ranging and complex, will affect all aspects of our lives, and will touch all people—some more so than others. Energy scarcity will result primarily in reduced capacity, capabilities, and services in the health care and public health systems. Climate change will cause new and increased demands on our current capabilities and services. Without preparation, early responses to these challenges will likely be motivated simply by rising and volatile energy prices—and characterized by trial and error, incorrect decisions, and highly politicized debate. Fortunately, we can plan ahead to provide communities with the essential capabilities and resources they'll need to be resilient, safeguarding individual and family health in an increasingly uncertain future.

## HEALTH AND ITS MANY DETERMINANTS

Health is a state of complete physical, mental and social well-being and not merely the absence of disease or infirmity.
—World Health Organization, 1948

### Physical, Mental, and Social Well-Being

A definition of health as merely the absence of disease is much too limiting. The broad

Note: this chapter has been excerpted from the original publication.

definition of health above was formulated more than sixty years ago and has remained widely useful. Although *physical* health is important to well-being, humans also need *mental* health, which starts with the absence of mental illness but also includes such concepts as freedom from fear of personal harm, freedom from fear about not meeting basic needs (food, water, shelter, safety), and so on. In addition, we are *social* creatures and require a sense of community: stimulating, trusting, and regular interactions with others, plus a sense of usefulness, satisfaction, and security in what we do and how we live our lives as members of groups. Without all three kinds of wellbeing—physical, mental, and social—we are not healthy.

When health and well-being have been defined broadly, it becomes easier to understand what the health impacts of energy scarcity and climate change are likely to be. Many things that are not considered to be "health related" *per se* are nevertheless important determinants of health; all of the factors below contribute to physical, mental, and social well-being—or lack thereof—and each of these is likely to be influenced by the coming energy and climate challenges:

- Community economic vitality
- Employment rate
- Social stability
- Neighborliness
- Dependability and affordability of basic needs like food and water
- Urban planning and design
- Reliable transportation systems
- Discrimination
- Political/military conflict
- Population dislocation/mass migration
- Confidence/worry about the future
- Equity/inequity
- Freedom of/restrictions on movement

- Disaster preparedness (how communities respond to droughts, floods, and heat waves)
- Availability of public health and health care services

Researchers studying what affects the health of society as a whole (as well as the health of smaller social groups) have identified "large-scale" factors far outside of an individual's control that can have profound influences on health.[1] For example, while tobacco smoking, inactivity, high-fat diets, high-salt diets, and obesity account for a large share of the world's incidence of heart disease, there is also substantial risk associated with adverse socioeconomic conditions. Thus, while many of us acknowledge that our behaviors influence our health in good ways and bad, few recognize the social context in which health-related behaviors occur and become socially patterned.[2] These social determinants of health are critically important and go a long way toward explaining different rates of disease across populations.

## Risk Regulators

Human actions and behaviors are influenced by many different dynamics and at many different scales. Here are just a few:

| Dynamics | Scales |
|---|---|
| Cultural | Global |
| Economic | National |
| Environmental | Community |
| Historical | Workplace |
| Political | School |
| Religious | Family |
| Social | Individual |

The opportunities and constraints created by these dynamics are called *risk regulators*, defined as features and phenomena of the social and built environments that shape, channel, motivate, and induce behavioral risk factors for adverse or good health outcomes.[3] For example, the ongoing war in the Congo that began in 1998 has been responsible for more than 5 million civilian deaths through 2009. Many of these persons did not die of war *per se*, but rather from the collapse of social and environmental life-support systems such as housing, public health, food, and water. The conditions created by the war exacerbated risks to public health beyond the immediate effects of the war itself.

The health threats of climate change and energy scarcity can be thought of in the same way. Climate change and energy scarcity have direct effects on health, such as through extreme weather events and the impacts of rising fuel prices on dietary choices. But they will also increasingly affect other systems that are essential to public health around the world, from food and water to economic activity and political stability. In addition, they will change the advantages and disadvantages different populations have for dealing with these crises. Some people and communities will find new opportunities in the changes brought by climate change and energy scarcity, but many more will be confronted with new and greater constraints. Most especially, populations that are already starting out disadvantaged in terms of resources, health, and political power—or with less of what might be termed *resilience*—will be more severely impacted.

## Health at Risk

One of the most important determinants of risk to our health in the United States has turned out to be the spatial organization of our communities. We have spent the last sixty years building a physical infrastructure—including highways, office buildings, housing subdivisions,

and shopping malls—that was entirely shaped by the availability of cheap and plentiful oil. Homes are far removed from jobs, services (including health services), and even places for recreation and social gathering—all things we need for our well-being. Thus our built environment becomes an important health risk regulator as energy scarcity makes distance more of an obstacle.

Another unexpected determinant of health risk is what might be termed our "provisioning system"—that is, the ways in which we provide our communities with the goods they need. The manufacture and transport of most goods will be impacted in obvious ways by the challenges ahead, but the health risk of the food system is probably the most worrisome. Our food generally comes to us from industrial models of food production, thousands of miles away and completely dependent on fuel, pesticides, herbicides, fertilizers, and plastics made from petroleum and natural gas—a very vulnerable situation in a future of oil prices double or even triple what they are today. In addition, climate change threatens to bring not just more crop-damaging extreme weather (especially droughts/floods) but major shifts in agricultural zones and pest ranges.

Climate change and energy scarcity will create direct challenges for our health system, but they will also create myriad *indirect* problems for health simply because of how we have built and provisioned our communities and economies up until now.

## Sustainable Well-Being

Dr. John Holdren, currently the lead science and technology policy adviser in the Obama administration, has argued that human well-being rests on a foundation of three pillars of conditions and processes[4]:

- *Economic:* production, employment, income, wealth, markets, and trade.
- *Sociopolitical:* national and personal security, liberty, justice, law, education, health care, science, arts, civil society, and culture.
- *Environmental:* air, water, soils, mineral resources, living organisms, and climate.

For sustainable human well-being, each is indispensable, none can be identified as the most important, and the three are highly interdependent. What Holdren did not make explicit was that energy scarcity and climate change will adversely influence each of the three pillars and limit options for responding to the tremendous challenges they create. He highlighted the "energy-economy-environment dilemma" that reliable and affordable energy is essential for meeting human needs, but the way we currently use energy is responsible for many challenging environmental problems. According to Holdren, "energy is the hardest part of the environment problem; environment is the hardest part of the energy problem; and resolving the energy-economy-environment dilemma is the hardest part of the challenge of sustainable well-being for industrial and developing countries alike."

Recent U.S. national-security reports state that climate change will pose serious threats to national security because it will likely increase poverty, lead to serious environmental degradation, and weaken national governments—findings that are interestingly similar to Holdren's three pillars and the idea of risk regulation.[5] A growing number of analysts are viewing climate change and energy scarcity through these lenses and the conclusion is unmistakable: Climate change and energy scarcity pose unprecedented challenges to human health and well-being.

# HEALTH CARE AND PUBLIC HEALTH

## Philosophy and Practice

Most people are familiar with the *health care system*—its primary function is to take care of us when we get sick, and as such is often referred to as the "illness care system." Your doctor's office, clinics, hospitals, medical laboratories, outpatient surgical facilities, and diagnostic centers such as for MRIs are all part of the health care system. It focuses on the health of an *individual* who seeks the advice of a health care practitioner for a specific health problem. The practitioner then typically takes a history, completes a physical examination, performs some laboratory or other diagnostic tests, formulates a diagnosis, and then offers a treatment plan. While the health care system usually focuses on persons who are sick, it also offers clinical preventive services to detect disease early (e.g., colonoscopy for colon cancer, mammogram for breast cancer) or prevent it (e.g., immunization for influenza).

In contrast, people are generally less familiar with the *public health system* and how it differs from the health care system. The goal of the public health system is to improve and maintain well-being in *communities and populations* rather than in individuals. In the United States it is made up of local and state health departments, public health laboratories, and the Centers for Disease Control and Prevention (CDC), all staffed with specially trained public health professionals. The public health system is most visible in its work to detect and stop outbreaks of infectious diseases (see box on page 1), but its functions also include:

- Working with government agencies to monitor air and water quality.

- Preparing for and providing services after a disaster (e.g., the provision of food, water, and shelter for affected residents).
- Enforcing health laws, such as by conducting restaurant inspections, to ensure health and well-being.
- Educating communities about good health-maintenance practices and informing the public about potential health hazards.
- Detecting, tracking, and responding to disease outbreaks.
- Mobilizing community members and organizations for health-related activities.
- Researching innovative solutions to health problems.

These are all useful functions not only for dealing with the consequences of climate change and energy scarcity, but also for working with the public to develop strategies for stabilizing the climate and preventing some of the negative consequences of energy scarcity.

## Challenges Facing the Health Care System

We have seen the impact of acute shortages on the health care system in patient transport via ambulance. After Hurricanes Gustav and Ike compromised refinery and transport capacity in the Gulf, the Southeast experienced a short-term fuel shortage, and in Atlanta our ambulances had to travel much farther than usual to fuel up. While service was maintained, this demonstrated for us the need to consider fuel storage for EMS organizations to maintain supplies in the face of acute shortages. It's the long-term price increases that are ultimately a greater concern,

however, as they will ramify throughout the health care system and are likely to cause significant inflation in health care costs above and beyond what we're already seeing from other sources. It's a strain that our health care system likely cannot bear.
—Dr. Jeremy Hess, Emergency Room Physician[6]

Hospitals are energy intensive, requiring high-technology diagnostic and patient-care equipment— and personnel to operate them— around the clock. Furthermore, concerns about spreading infection have led to the reliance on disposable single-use supplies and equipment, which contributes to the resource use of hospitals. Rising energy costs will add a significant burden to the future cost of providing illness care.

One of the most energy-intensive aspects of the health care system is emergency medical assistance. Current practice guidelines recommend that only the level of care required to stabilize the patient be provided in the field. The patient is then transported to a hospital emergency room or tertiary care provider such as a burn center or a trauma center to receive the majority of his or her medical care. Emergency transportation typically occurs in helicopters, with an average fuel mileage rate of about 10 miles per gallon (range is 1 to 15 miles per gallon, depending on flying conditions and the size and power of the helicopter), or ambulances, with an average fuel mileage rate usually less than 10 miles per gallon (although newer designs can approach 20 miles per gallon). Emergency medical transport companies faced economic hardship during the summer of 2008 when gasoline exceeded $4 per gallon. With petroleum prices increasing, the long-term feasibility of this system as currently configured is questionable.

## SARS and the Public Health System

The events surrounding the severe acute respiratory syndrome (SARS) epidemic of 2002–2003 provide a good example of the public health system in action, and also a good example of what can happen if it doesn't work well. The SARS epidemic began in Guangdong Province, China, in late 2002. For a variety of reasons, the response of the Chinese public health system was not adequate, and SARS spread rapidly, eventually infecting people in thirty-seven countries. Once the World Health Organization learned what was going on, however, public health professionals sprang into action to research and identify the cause of the illness, determine how it was transmitted and how to stop its spread, and inform public health agencies around the world. Here in the United States, monitoring and surveillance systems were set up to detect the first cases, make sure adequate treatment was obtained, and prevent the spread of the disease to others by tracing contacts and keeping infected persons separated from others. What could have become a global epidemic was reined in and successfully stopped by the public health system.

Petroleum-based transportation of patients is also important to the health care system in non-emergency situations. Patients must typically transport themselves to clinics and hospitals to receive care, medical personnel must transport themselves to work every day, and patients are transported from one medical facility to another for specialty medical care or to be closer to family. Many rural areas are already experiencing what suburbs and urban areas might expect to see in a future of energy scarcity, with more difficult access to health care and more cost associated with transit to and from the hospital for routine or urgent care.

Petroleum is a basic manufacturing material for many medications and therefore its increasing scarcity could well make medications more expensive. Many people will likely be willing to bear the added cost for necessary medications; patients on fixed incomes, however, will find it more difficult to pay for needed prescriptions and their health will suffer accordingly. Petroleum is also a basic material in many medical supplies, especially those made of plastic such as intravenous bags and tubing, syringes,

and oxygen masks. Plant-based alternatives may provide acceptable substitutes but will require testing for potential negative interactions with pharmaceuticals and body fluids.

## Challenges Facing the Public health system

> The connections among the global just-in-time economy, energy availability, and public health are far more extensive than almost anyone can imagine.... [T]he public health community has been largely absent from this consideration and discussion of energy issues.
>
> —Dr. Michael Osterholm, director of the Center for Infectious Disease Research and Policy (CIDRAP) at the University of Minnesota[7]

The public health system will face unprecedented challenges from energy scarcity and climate change. Threats to population health and the public health system's inability to deal with them will be even greater challenges than the already formidable issues facing individual health care.

The backbone of the public health system is the network of health departments in every state and locality. Energy scarcity will likely be felt first by these agencies in their budgets; transportation costs will be the first spike, but upstream transportation costs involved in providing materials and supplies will also rise, stressing already tight budgets in practically every area. Eventually, services will have to be cut and/or models of delivery redesigned.

Many communities are located in areas at risk from extreme weather, seismic or volcanic activity, sea-level rise, and freshwater floods. When such risks become actual events, immediate disaster response usually involves the rapid transport of massive amounts of medical supplies and personnel into an affected area. Disaster preparation for some events also involves the rapid evacuation of the population. With transportation costs rising as a result of energy scarcity, however, disaster planners will need to design new models for preparation and response.

Transportation is an integral component of the public health system. Although some direct services rely on individuals going to a clinic or office, many local and state health departments provide services that require department personnel to travel regularly. For example:

- Food inspectors must travel around to the places where food is stored, prepared, and sold.
- Outreach workers must travel to the homes of tuberculosis patients for many months to observe them taking their medicine—even after their symptoms are gone—so as to avoid spread of the disease.

- Investigations of disease outbreaks require health professionals to gather information from many sources in the community to ascertain the cause of the outbreak and devise a plan to stop it.

## CHALLENGES FOR HEALTH IN GENERAL

Until recently, most public discussion of the adverse health impacts of fossil fuels has focused on their combustion (i.e., air pollution) rather than on the overwhelming fossil-fuel dependence of the systems that influence our health.[8] This discussion needs to change, as discussed above, and it also needs to broaden to include climate change in particular and the interaction of energy scarcity with environmental challenges in general. The problem is that we are entering the era of energy scarcity at the same time that the effects of climate change, ecosystem degradation, and species and biodiversity losses are accelerating.

For more than a century, we have used cheap and plentiful energy to insulate ourselves from the negative health consequences of our environmental destruction. If we depleted fish stocks in one area, we trawled deeper and farther using cheap energy to harvest other species. If we degraded ecosystem services such as capturing, purifying, and storing freshwater, we used cheap energy to drill deeper into aquifers or built desalination plants, a direct way of converting energy into potable water. If drought adversely affected food production in one locale, we used cheap energy to transport food great distances from elsewhere. If severe storms impacted our cities, we used cheap energy to bring in relief supplies, rebuild, and, in some cases, put in place structures to lessen the effects of the next storm.

We will soon no longer be able to use cheap and plentiful energy to mask the effects of the massive environmental changes we've caused.

Global climate change is occurring and if left unchecked will have severe consequences for the health and well-being of citizens of every nation. Even under the best-case scenarios, an average global temperature increase of 2 degrees Celsius (3.6 degrees Fahrenheit) above pre-industrial averages is inevitable. Even that seemingly small amount of average global warming will have significant health impacts.

Finally, energy scarcity and climate change have important implications for health disparities—that is, large differences in rates of disease for populations that differ by race/ethnicity or socioeconomic status. These concerns are already a key challenge for public health but are likely to get much worse as populations face the local effects of climate change, increasing or volatile energy prices, and international population movements forced by these challenges. Local movements of people, including wealthier, previously suburban-dwelling families returning to cities and displacing predominantly poorer minority families, will also influence population health.

## Heat Stress and Worsening Air Quality

Global warming is tracked by following the average global temperature, but averages can be misleading. For example, relatively small average temperature increases mask one of the hallmarks of climate change: more frequent and longer-lasting severe heat waves. In 1995, a heat wave hit Chicago resulting in more than 700 deaths; more than 45,000 people died in heat waves during the summer of 2003 in Western Europe; and the summer of 2006 brought scorching heat to much of the United States and Canada, killing 300 in California alone and sending tens of thousands to emergency rooms and hospitals.[9]

Some people are more vulnerable to heat than others, including babies, children, the elderly, the poor, those who live in inner-city neighborhoods, and the socially isolated (again highlighting the importance of social well-being). More Americans die every year from heat stress than from any other weather-related event, with the exception of Hurricane Katrina. Computer models suggest that if climate change occurs unabated ("climate chaos"), by 2040 heat waves as severe as the 2003 event that killed so many people in Europe could occur every other year.[10]

Heat waves are especially deadly because warmer temperatures worsen air quality. For example, higher temperatures increase concentrations of ground-level ozone (the primary component of smog and an important contributor to global warming), which damages the lungs, blood vessels, and heart. People who have asthma and other breathing disorders are especially vulnerable to the effects of ozone, requiring more medications and leading to more emergency-room visits and hospitalizations. The combination of high temperatures and high ozone concentrations is especially deadly and plays an important role in the numbers of people who die during heat waves.[11] Other kinds of air pollution are expected to get worse with climate change as well.

Air conditioning is a partial and temporary fix for heat waves but requires substantial amounts of electricity. With the majority of the American electricity supply coming from burning coal, the use of air conditioning over the long term will only exacerbate the climate problem. Making matters worse, power plants themselves become overheated during heat waves and are sometimes forced to shut down, as happened in Greece during heat waves in 2007 and in the U.S. Southeast during the 2006–2008 drought. Energy production is also expected to be constrained by climate change because of impacts on local precipitation cycles, reducing river flows. We simply cannot solve our climate-related health challenges by increasing our energy use.

## Infectious Diseases

Warmer temperatures, milder winters, precipitation changes, and other effects of climate change can influence the distribution and risk of many infectious diseases. Debilitating and deadly insect-borne diseases such as malaria, dengue fever, and Lyme disease are especially sensitive to changes in temperature, humidity, and rainfall patterns and will likely increase their ranges and possibly their transmissibility (see box on page 170). Waterborne infectious diseases will also be influenced by warmer temperatures, changes in precipitation patterns, and the compromised ability of degrading ecosystems and suboptimal built environments to deal with heavy precipitation events.

## Threatened Water and Food Supplies

Clean water is vital to life and health, but climate change will seriously threaten water supplies around the world. In general, regions that are typically wet now will get wetter and those that are dry now will likely get drier. Even wetter regions, however, may still experience frequent bouts of water shortages because more of the precipitation will come in the form of heavy precipitation events, leading to greater runoff of stormwater and longer periods of droughts in between rain events.[12]

Much of the world's population gets its freshwater from glaciers and mountain snowpack. Mountain ranges collect water, purify it, store it as ice or snow, and release it over time into streams and rivers. Global warming has already caused many of these glaciers and snowpacks to melt far more rapidly than expected. If glaciers in the Andes continue melting at present rates, they will likely disappear completely within one

or two decades. The Himalayas are melting so fast that Chinese and Indian farmers are seeing more river water than usual, making it even harder to motivate the water-conserving changes necessary as these glaciers—which provide freshwater for more than one-sixth of the world's population—disappear. Much of the western United States is in a similar situation, relying on the threatened snowpack of the Rocky Mountains, the Sierra Nevadas, and the Cascade Range for much of its water. This has serious implications for the entire country: The farms of California alone account for approximately half of U.S.-grown fruits, nuts, and vegetables and more than one-fifth of the milk supply.[13]

There are many reasons why the effects of climate change could lead to diminished food supplies. Plants require certain amounts of moisture and nutrients and can live only in particular temperature ranges. With climate change, previously productive agricultural zones are becoming too hot or too dry to grow some traditional crops, forcing the production of those crops northward or eliminating their cultivation altogether. For example, by the end of the century the climate of New Hampshire is projected to be like that of North Carolina today.[14] Unfortunately, the plant and animal species in New Hampshire evolved over hundreds and thousands of years to take advantage of the climate of northern New England, not the climate 800 miles to the southwest.

Other negative climate effects on agriculture include:

- More frequent, longer-lasting, and more severe droughts and floods
- Warmer temperatures
- Chronic water shortages
- Higher concentrations of ozone and other air pollutants that hamper plant growth
- Stronger and more resistant plant pests and diseases

---

### Climate Change and Dengue fever

*Bill McKibben, as quoted in "Americans Who Tell the Truth"*

I wrote the first book [*The End of Nature*] on climate change in 1989, so i've been writing and speaking about it for a long time. But some years ago i took a trip to Bangladesh to do some reporting. And while I was there they were having their first big outbreak of dengue fever, a mosquito-borne disease now spreading like wildfire because of global warming. Since i was spending a lot of time in the slums, i got bit by the wrong mosquito myself.

I was as sick as i've ever been, but because i was healthy going in, i didn't die. Lots of people did, mostly old and young. I remember standing in the hospital ward looking at rows of cots of shivering patients, and thinking to myself: these people did nothing to deserve this.

You can barely measure how much carbon Bangladesh produces: a nation of 140 million, but most of them don't have electricity or cars. Whereas in the U.S. 4% of us manage to produce 25% of the world's carbon dioxide. A quarter of those hospital beds were our fault.

When I figured that out in my gut, i came home and started organizing, and that's what i've been up to ever since.

Source: Americans Who Tell the Truth, "Bill McKibben," http://www. americanswhotellthetruth.org/pgs/portraits/bill_mckibben.php.

---

Energy scarcity promises to further compound our food-production challenges. The American food supply has become reliant on fewer and larger farms using a number of fossil-fuel inputs in the form of fertilizers, pesticides, pump irrigation, heavy machinery, and long-distance transport to accomplish what was once done by many people, locally, without the various fossil-fuel inputs. Constant erosion of topsoil and cultivation of single-plant crops have led to a reliance on chemical fertilizers created from natural gas and chemical pesticides largely derived from petroleum. Practically all communities now rely on food produced using these methods and transported long distances—frequently from other regions of the United States, other countries, or even from the other side of the world—to arrive at our grocery stores. Without a steady supply of inexpensive fossil fuels, especially petroleum, the current American food system will not

function—and health depends on an adequate supply of nutritious food.

## Extreme Weather Events

The most well-known effect of climate change is rising sea levels, which threaten coastal areas with inundation. Climate change also promises to bring more severe and potentially more frequent extreme weather events such as hurricanes, tornadoes, and heavy rainfall, all of which increase the risk of injury and death and cause social disruption (itself an important cause of adverse health impacts). With more of our precipitation coming in less-frequent heavy-rainfall events, freshwater flooding is becoming a greater problem, creating both immediate and long-term risks. A recent study documented that the majority of outbreaks of waterborne disease in the United States occurring between 1948 and 1994 followed heavy-rainfall events

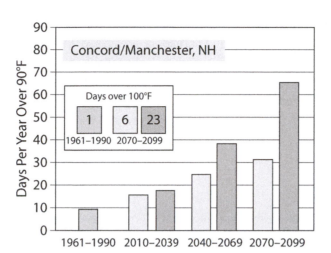

FIGURE 1. BY THE END OF THE CENTURY THE CLIMATE OF NEW HAMPSHIRE MAY BE LIKE THE CLIMATE OF NORTH CAROLINA TODAY.

that overwhelmed water treatment facilities.[15] Another study found that flood survivors often experience greater risk of chronic diseases such as diabetes and heart disease long after the floodwaters have subsided.[16] For decades, evidence has mounted that centuries of building levees and draining wetlands has increased the risk of flooding—another example of how the decisions we make about our communities and economies can indirectly but profoundly affect public health risks.

Other evidence of how failed ecosystems wreak havoc on human settlements, compromising health and wellbeing, includes the finding that intact coastal wetland ecosystems might have reduced the height and/or speed of the storm surge created by Hurricane Katrina in 2005 and possibly prevented the resultant flooding that devastated New Orleans.[17] The loss of these coastal wetland ecosystems is largely due to human activities such as offshore oil drilling and refining, importation of invasive species, and boxing in the Mississippi River so that it is no longer able to deposit sediment to replace coastal land lost to erosion.

## Conflict and Health

Competition for shrinking environmental resources, especially the necessities of water, food, and housing, could potentially result in greater conflict within and between geopolitical entities. There is ample evidence that the scarcity of environmental resources has played an important role in many areas of conflict, such as the genocides that devastated Rwanda and continue to occur in Darfur, the ongoing clashes between Zapatista rebels and the federal government in Mexico, and the decades-long modern Israeli-Palestinian conflict.[18] That is, what have been termed "ethnic conflicts" have actually been exacerbated by, if not directly caused by, environmental scarcity. Such conflicts will increase in the era of energy scarcity and climate change. In addition, the number of environmental refugees created by rising sea levels and failure of the local ecosystems to meet basic needs could increase by many hundreds of millions.[19] These refugees will face a greater risk of attack and conflict if they must cross political or cultural borders and will face the same hardships in many countries as those who flee war zones. As climate change worsens the gap between those with resources and those without, social unrest may worsen and spread into previously stable areas. This is another example of a risk regulator. If left unchecked, environmental degradation and the challenges it creates can ultimately threaten the basis of society itself.

## Mental Health Effects

The risks to mental well-being in a future of energy scarcity and climate change are quite significant. Examples include persons forced from their homes due to extreme weather events; the inability of the environment to provide sufficient food and water; individuals faced with job loss, separation from family and friends, and concern about the future; and persons coping with the various disruptions to life caused by an unstable climate. Such mental health outcomes as depression, anxiety, and post-traumatic stress disorder are expected to increase as a result.[20]

## THE FUTURE OF HEALTH CARE AND PUBLIC HEALTH

What health care and public health services will look like in a world constrained by energy scarcity and climate change depends on the choices we make in response to these challenges—and what happens to economies and communities as a result.

Energy prices will influence where people live and how they transport themselves from place to place. It is likely that these changes will influence future communities to be more compact, higher density, and more walkable; to contain a mixture of residential, commercial, and even recreational uses; and to be transit accessible. At the same time, populations will be on the move from the effects of climate change,

including extreme weather events, drought, sea-level rise, declining food production in some areas, and the inability to make environments hospitable owing to energy costs. To meet the needs of an energy-constrained future, the public health system as well as the health care system will need to be redesigned.

Unfortunately, policy discussions about reforming the health care system have not mentioned energy scarcity or climate change. The next iteration of our health care system must nevertheless be designed to meet these challenges and function within their constraints. In general, the allocation of financial resources, personnel, and energy should be shifted to favor public health and preventive services—for all members of society—instead of focusing primarily on illness care. More than anything else, this would result in lower energy requirements for the health care system while building more resilient communities nationwide, benefiting all citizens at all socioeconomic levels.

There are many other opportunities to improve health care and public health while preparing for the energy- and climate-constrained future. Of these, the three largest have to do with delivery of services, disaster response, and food.

## Delivery of Services

The most immediate challenge to health in the United States is the sheer energy intensity of our current health care system. We present four basic ways in which this challenge could be met. In reality, a combination of many approaches will likely be needed.

1. *Reorganize where care is provided.* For many years the trend has been to centralize health services in larger hospitals and reduce services in or even close smaller hospitals. Care for rare or complicated illnesses will need to remain in centralized high-level

hospitals, but smaller hospitals and care centers that could care for most medical problems could be more numerous and located at midlevel transportation hubs for easier access. Health services could even be brought directly into neighborhoods, through many small, scattered, local "pods" of public health workers who are able to perform many functions in their specific neighborhoods. Many of these changes would require reworking the way public health professionals are educated and trained.

2. *Use technology to decentralize more services.* Remote imaging and video conferencing technologies, which allow specialists at larger institutions to diagnose and treat patients remotely, can be used to improve care at decentralized facilities. These technologies can also be used to develop and monitor skills in localities and to transfer skills there. Of course, such high-tech solutions would require energy and resources to put into place and to maintain. As transportation becomes more expensive, new models of health service delivery will be required.

3. *Practice medicine differently.* Some energy-intensive procedures may not be medically necessary. We recommend reducing reliance on energy-intensive diagnostic procedures when those procedures are used only to confirm what a physician knows from a physical exam and less invasive tests. This would require changes to how health care providers are reimbursed and might also require a change in the liability laws so that health providers do not feel obligated to defend their clinical judgment from potential lawsuits with unnecessary diagnostic tests.

4. *Plan for smart decision-making about energy usage.* As fossil-fuel supplies decline, the costs of transporting goods and people,

powering facilities and machines, and manufacturing synthetic products will all rise. The health care industry should follow the example of most other industries and push for greater energy efficiency in its facilities and technologies. It may also be necessary to consider how the increasingly scarce resources will be apportioned for energy use (e.g., emergency medical transport) versus materials use (e.g., pharmaceuticals, plastics).

## Disaster Response

Disaster responses currently rely on the ability to ship in resources and personnel from afar and to evacuate large populations if necessary. As energy prices rise, disaster preparedness will need to focus far more on building local resources and training local personnel. Emphasis should shift from disaster response to disaster prevention and preparedness based on a community's specific areas of vulnerability. Typical efforts to reduce future impacts of disasters, such as building more and taller seawalls or drainage systems, may also be limited by energy scarcity. Alternative—and often smarter—approaches will be required, such as changing zoning regulations to disallow additional development in high flood-risk areas and using native vegetation along wetlands and waterways to buffer storm surges and reduce erosion.

## Food

Like most other industrialized countries, the United States will need to redesign its methods of producing, importing, and distributing food. Our eating habits will need to change. No longer will much of the nation be able to eat fresh fruits and vegetables flown in from California, South America, and elsewhere in the dead of winter. Instead, during natural harvest times, produce will have to be canned, dried, or otherwise preserved for consumption during the off-season. The public health system can play an essential role in building local food resilience by supporting and educating about:

- Localized food production and processing
- Farmers' markets and regional purchasing by groceries and supermarkets
- Gardening and farming projects in schools and by community groups
- Regional and seasonal cuisine
- Safe food handling and food preservation (canning, drying, etc.)

One significant benefit of increasing local food production is a greater supply of fresh food (if only seasonally for some locations), which often is less processed, better tasting, and more nutritious. Food production may ultimately need to occupy a greater proportion of time for many individuals than has become the norm over the last hundred years. Yet another potential benefit of this is the building of social capital as communities work together to grow and process food, securing a mutually shared food future.

## HOW WE CAN ADAPT

The effects of climate change will create new demands on our health care system and for public health services. We must prepare for this reality, while also doing everything we can to reduce our contributions to global warming. All the health system adaptations we can envision and all the ways to enhance community resilience that we can call for—much less implement—will not be enough if the climate is not ultimately stabilized.

All forms of energy, other than passive warming from the sun, have an environmental

and a societal cost. Therefore, using less energy or using it more efficiently should be a primary societal goal, regardless of where that energy comes from. To accomplish this goal, housing patterns, transportation options, food and water provisioning, and many other aspects of our lives will all have to be redesigned to require substantially less energy from any source. Energy scarcity will force these decisions upon us, but hasty decisions to replace petroleum with other liquid fuels, such as ethanol, biodiesel, or oil from oil sands, will only forestall the inevitable for a short time and will greatly aggravate other problems, such as climate change and food and water insecurity.

Some options, however, for addressing the dual challenges of climate change and energy scarcity could make our communities better places to live. A stronger sense of community, greater emphasis on family and friends, less time spent in cars and commuting, and localization of economic activity and food production will all benefit health and well-being.

Transforming our health care and public health systems will require significant policy changes. It's essential that citizens educate their elected officials about the issues and demand prompt, well-informed, forward-looking solutions. This will not be easy, because the necessary changes will likely be seen as politically unpopular and volatile energy prices will encourage actions that do not necessarily serve society well in the long run. But if we make the right choices now, we can maximize the benefits and lessen the risks. The transition to the energy-scarce, climate-constrained future will create significant hardship if tough decisions about how to proceed are not made soon. However, the end result of a more self-sufficient, cohesive, resilient, and healthy society is worth the effort.

## NOTES

1 Thomas A. Glass and Matthew J. McAtee, "Behavioral Science at the Crossroads in Public Health: Extending Horizons, Envisioning the Future," *Social Science and Medicine* 62, no. 7 (2006), 1650–1671.

2 In the public health field, "social patterning" is the idea that differences in states of health across populations are determined by differences in the distribution of a variety of advantages and disadvantages in those populations.

3 Glass and McAtee, "Behavioral Science at the Crossroads in Public Health."

4 John P. Holdren, "Science and Technology for Sustainable Well-Being," *Science* 319, no. 5862 (January 25, 2008), 424–434.

5 House Select Committee on Energy Independence and Global Warming of the House Permanent Select Committee on Intelligence, *National Intelligence Assessment on the National Security Implications of Global Climate Change to 2030*, statement for the record of Dr. Thomas Fingar, deputy director of National Intelligence for Analysis and chairman of the National Intelligence Council, June 25, 2008.

6 Dr. Jeremy Hess, personal communication with the authors, February 17, 2010.

7 M.T. Osterholm and N.S. Kelley, "Energy and the public's health: making the connection," *Public Health Rep* 124, no. 1 (2009), 20–21.

8 Charles A. S. Hall and John W. Day, "Revisiting the Limits to Growth after Peak Oil," *American Scientist* 97, no. 3 (May–June 2009), 230–237.

9 Jan C. Semenza et al., "Heat-Related Deaths during the July 1995 Heat Wave in Chicago," *New England Journal of Medicine* 335, no. 2 (July 11, 1996), 84–90; Tom Kosatsky, "The 2003 European Heat Waves," *Eurosurveillance* 10, no. 7 (July 1, 2005), 148–149; Kim Knowlton et al., "The 2006 California Heat Wave: Impacts on Hospitalizations and Emergency Department Visits," *Environmental Health Perspectives* 117, no. 1 (January 2009), 61–67; Bart D. Ostro et al., "Estimating the Mortality Effect of the July 2006 California Heat Wave," *Environmental Research* 109, no. 5 (July 2009), 614–619.

10 P. A. Stott, D. A. Stone, and M. R. Allen, "Human Contribution to the European Heatwave of 2003," *Nature* 432 (December 2, 2004), 610–614.

11 Mercedes Medina-Ramon and Joel Schwartz, "Temperature, Temperature Extremes, and Mortality: A Study of Acclimatisation and Effect Modification in 50 United States Cities," *Occupational and Environmental Medicine* 64 (2007), 827–863.

12 Anil V. Kulkarni et al., "Glacial Retreat in Himalaya Using Indian Remote Sensing Satellite Data," *Current Science* 92, no. 1 (January 10, 2007), 69–74.

13 California Department of Food and Agriculture, "Agricultural Statistical Review," in *California Agricultural Resource Directory 2007* (2008).

14 U.S. Global Change Research Program, *Our Changing Planet: The U.S. Climate Change Science Program for Fiscal Year 2009* (Washington DC: U.S. Global Change Research Program, 2008).

15 Frank C. Curriero et al., "The Association between Extreme Precipitation and Waterborne Disease Outbreaks in the United States, 1948–1994," *American Journal of Public Health* 91, no. 8 (2001), 1194–1199.

16 Mike Ahern et al., "Global Health Impacts of Floods: Epidemiologic Evidence," *Epidemiologic Reviews* 27, no. 1 (2005), 36–46.

17 Pat J. Fitzpatrick et al., "The Impact of Louisiana's Levees and Wetlands on Katrina's Storm Surge," presented at the 28th Conference on Hurricanes and Tropical Meteorology, American Meteorological Society, May 2008, http://ams.confex.com/ams/28Hurricanes/techprogram/paper_137224.htm.

18 Jared Diamond, *Collapse: How Societies Choose to Fail or Succeed* (New York: Penguin, 2005); Thomas F. Homer-Dixon, *Environment, Scarcity, and Violence* (Princeton, NJ: Princeton University Press, 2001).

19 Intergovernmental Panel on Climate Change (IPCC), "Summary for Policymakers," in *Climate Change 2007: Impacts, Adaptation and Vulnerability*, M. L. Parry et al., eds. (Cambridge, UK: Cambridge University Press, 2007), 7–22.

20 A. E. Kazdin, "Psychological Science's Contributions to a Sustainable Environment: Extending Our Reach to a Grand Challenge of Society," *American Psychologist* 64, no. 5 (July– August 2009), 339–356; Robert Gifford, "Psychology's Essential Role in Alleviating the Impacts of Climate Change," *Canadian Psychology* 49, no. 4 (November 2008), 273–280; R. C. Kessler et al., "Trends in Mental illness and Suicidality after Hurricane Katrina," *Molecular Psychiatry* 13, no. 4 (April 2008), 374–384.

# Section VI

# COMMUNITIES AND THE BUILT ENVIRONMENT

# ENVISIONING SUSTAINABLE PLACES

By Tim Beatley and Kristy Manning

How do concepts of sustainability translate into a vision and plan for sustainable places? What is a sustainable place? What does one look like, and would we know one if we saw it? What would be the distinguishing characteristics of such places? What follows is an initial exploration of these questions and an attempt to visualize what sustainability can mean at community and regional levels.

## SUSTAINABLE PLACES ACKNOWLEDGE FUNDAMENTAL ECOLOGICAL LIMITS

Whether they are called "green communities," "green cities," or "ecocities," sustainable places seek to limit environmental impacts and the consumption of natural resources. Cities and urban developments have tremendous ecological impacts, and the seriousness of the environmental crisis to which they contribute suggests the need for a fundamentally new governance and management approach—one that acknowledges and implements a new ecological paradigm. As Mark Roseland observes in *Toward Sustainable Communities*, "Cities provide enormous, untapped opportunities to solve environmental challenges, and local governments must and can pioneer new approaches to sustainable development and urban management" (Roseland 1992, p. 22).

Concern for environmental protection and conservation are not new to the planning and design fields, and the creation of sustainable communities builds on these concerns. Important early examples include Ian McHarg's groundbreaking *Design with Nature* (1969) and the attention paid to carrying capacity, protection of sensitive lands, watershed planning, and ecological land planning (e.g., Schneider, Godschalk, and Axler 1978; Godschalk and Parker 1975; Spirn 1984; Steiner 1991; Marsh 1991; Thurow, Toner, and Erley 1975; Salvesen 1990). What is new, perhaps, is the commitment and priority given to respecting ecological limits in the planning, design, and operation of our communities. Planning for sustainable communities is not simply a matter of avoiding a few wetlands, or saving a few acres of open space, or putting in place a few nonpoint best-management practices. Rather, it requires considering ecological limits and environmental impacts at every step of community development and in every aspect of community design, from the energy efficiency of buildings to the regional transportation system to how the industrial and commercial sectors in the community go about business. Planning for sustainability means reorganizing the social, physical, and political-economic landscape in very fundamental ways.

What will guide communities in determining how far to go in reducing their environmental impacts? How will we know when ecological sustainability is reached? A good place to start, perhaps, is with local and regional indicators, and with a concerted effort not to exceed the natural earning capacity of regional-level ecosystems or bioregions (for a discussion of bioregion delineation, see Sale 1991). The notion offered by the National Commission on the Environment that we should be living off our ecological interest (as opposed to depleting natural capital) does have clear, although not uncomplicated, applicability to such actions as groundwater extraction, air and water contamination, use of agricultural and forest lands, and maintenance of biodiversity and habitat. Sustainable communities seek to live, develop, and operate within these natural "limits."

A sustainable community is a place that seeks to contain the extent of the urban "footprint" and strives to keep to a minimum the conversion of natural and open lands to urban and developed uses. The evils of urban sprawl are many and well documented (e.g., Kunstler 1993; Roseland 1992; Downs 1994). From an environmental or conservation point of view, loss of habitat and many other serious forms of environmental degradation find their roots in wasteful and destructive development patterns and in the economic and social system that encourages them. Increasingly, to have an effective environmental policy (whether the concern is with biodiversity loss, air quality, or coastal management) first requires an effective urban policy.

Sustainable communities, then, are places that exhibit a compact urban form. This is quite a challenge in the United States, where the trend has been in the other direction (Downs 1994).

The 1994 "Full House" study, for example, renewed significant concerns about preserving remaining farmlands while "satisfying the demand for residential building, industrial sites, parking lots, shopping centers, schools, and recreation areas" necessary to accommodate further population increases (Brown and Kane 1994, p. 166). In sustainable places, the conservation and protection of natural and undeveloped land is a primary goal that is not easily overruled by proposals for housing or other uses that could be accommodated within existing urban areas. A different physical form accompanies such a vision and includes compact development patterns, higher densities, and more land-efficient development projects; a more sensible and sustainable transportation system, with less reliance on automobiles, more availability of public transit, and more opportunities for walking and bicycling; greater emphasis on infill, adaptive reuse of buildings, and re-urbanization; protection and avoidance of ecologically sensitive lands; and avoidance and conservation of natural hazard areas, including flood plains, high-erosion zones, and areas subject to wildfires and landslides. Curtailing the consumption of land at the urban periphery becomes a critical part of preserving biodiversity, important ecological functions, and productive lands essential for sustaining current and future populations. Serious urban-growth containment, then, becomes a central part of preserving these essential natural and biological resources.

Sustainable places seek more broadly to minimize the extent of their "ecological footprint," to use William Rees's important concept. To provide food, housing, and energy for a community's population requires a tremendous drain on collective ecological capital—the equivalent of five hectares per year per North American resident. These consumption levels can be supported only by drawing upon the resources and carrying capacities of other regions and nations. Sustainable places seek to better understand these extra-local resource demands and to minimize the extent to which the carrying capacities and resources of other bioregions are "appropriated." A sustainable place, therefore, thinks beyond its own local and regional ecological limits and capacities. It seeks every possibility to reduce these demands whether through energy-efficient buildings and transportation, other sustainable building practices, use of renewable energy sources such as solar, or pricing policies that discourage waste.

This necessarily raises the question of whether sustainable places should aspire to being self-sufficient in terms of the resources they require and the environmental impacts they generate. This is not an easy question to answer, and there is no hard and firm standard. Much of the current sustainability literature identifies self-sufficiency as a desirable goal (e.g., Sale 1991). The problems arise when cities (and nations) exploit and degrade the environments of other regions and nations to satisfy wasteful and opulent patterns of development and consumption at home. Trade among cities, regions, and nations, however, is not itself a bad thing and is, in fact, central to the functioning of our global economy. Reducing the unnecessary import of resources and goods from other regions, as well as the export of wastes and pollutants, should be an important goal. Local food production is preferable, for example, because it results in lower energy consumption, less pollution, and healthier food. It may not be possible, for climatic or other reasons, to produce certain foods locally, however. Becoming self-sufficient in the sense of reducing unnecessary imports—things that are or could be produced locally—and minimizing or ceasing the export of wastes and pollutants, is a desirable goal.

## SUSTAINABLE PLACES ARE RESTORATIVE AND REGENERATIVE

Some definitions of sustainability imply preservation of the status quo—simply protecting or ensuring that the conditions of the present are not diminished in the future. But the ethical imperative of sustainability calls for more: it calls for energetic efforts to reverse the degradation already brought about and for passing along to our children and their children a planet where both the natural and built environments are of a higher quality and condition.

There are, unfortunately, reasons to be skeptical about our ability to meet these challenges, given current global trends of population growth, consumption, and environmental degradation. Yet at local and regional levels, the higher standards implied by sustainability are not only more conceivable, but necessary from a practical as well as an ethical standpoint. Indeed, there are encouraging examples of efforts to restore the environment, or at least important pieces of the environment. These range from recreating prairies in the Chicago area to reintroducing the red wolf in North Carolina to the reforestation efforts under way in a number of places (see Baldwin, DeLuce, and Pletsch, eds. 1993; for a good discussion of ecological restoration as a new perspective on the relationship between humans and nature, see Jordan 1993).

## SUSTAINABLE PLACES STRIVE FOR A HIGH QUALITY OF LIFE

To be a viable paradigm, sustainability must incorporate a strong social component. Along with ecological issues, then, sustainable communities are equally concerned with social and human sustainability—creating and supporting humane living environments, livable places, and communities that offer a high quality of life. Characterizing the social dimension becomes much vaguer and more subjective than defining the environmental or ecological dimension; we cannot rely as readily on such measures as natural carrying capacity. Terms such as livability and quality of life, in turn, beg for definition and description and, ultimately, may be up to the community itself to clarify.

It seems fairly clear, however, that current patterns of urban development in the United States are becoming more socially and psychologically stifling. We have created, or allowed to be created, urban and suburban places that will not stand the test of time. Increasingly, downtown urban areas are devoid of activity after 5:00 P.M.; meanwhile, suburban and edge cities lack charm and a sense of place and depend heavily on the automobile. The zealous separation of land uses promoted by conventional zoning and the scattered, sprawling development patterns that characterize much of our contemporary landscape require an increasingly dysfunctional transportation system that is plagued with traffic congestion and long work commutes (Fig. 2.1). People are increasingly isolated from one another and from a connection to the larger community.

The creation of vibrant and active urban spaces is surely an important part of making places more livable. Europe abounds with urban places that bubble with aliveness—streets such as Las Ramblas in Barcelona or Stroget in Copenhagen, for example (Jacobs 1993). The characteristics of these timeless communities are neither mysterious nor surprising: walkability, mixed land uses and activities, a density of people and commerce sufficient to create vital, active places of activity, and pedestrian and public spaces that encourage this activity. The successful renaissance of several U.S. cities—Baltimore is a good example—has recreated this kind of energy while at the same time increasing the efficiency of land use and encouraging greater use of public transportation.

Concern with the social dimensions of sustainability suggests that a sustainable community is one that addresses a host of related goals, including adequate and affordable shelter, health care, and other essential services to residents; a safe and crime-free environment; and humane and stimulating work environments. Marrying social and environmental concerns, then, is a major characteristic of a sustainable community.

## "PLACE" MATTERS IN SUSTAINABLE PLACES

"Sense of place" is also an important dimension of sustainability. Many Americans have little attachment to place largely because many of the places we build are not deserving of attachment, or because we have been quick to demolish those community features that have traditionally provided meaning in our everyday lives. Absent this feeling of attachment, it is little wonder that we are so quick to uproot ourselves when a new employment opportunity arises, the local school system fails to deliver, or a new ex- urban development promises "country living with all the amenities." Ultimately, this pattern of mobility feeds a vicious circle: the more we move around in search of the elusive "better" place to live, the less we invest in our communities, such that those few features that do provide a sense of attachment are being abandoned.

To foster a sense of place, communities must nurture built environment and settlement patterns that are uplifting, inspirational, and memorable, and that engender a special feeling of attachment and belonging. A sustainable community also nurtures a sense of place by understanding and respecting its bioregional context—its topography and natural setting, its rivers, hilltops, open lands, native flora and fauna, and the many other unique elements of its natural context. A sustainable community respects the history and character of those existing features that nurture a sense of attachment to, and familiarity with, place. Such "community landmarks" may be natural—a meadow or an ancient tree, an urban creek—or built—a civic monument, a local diner, an historic courthouse or clock tower. Finally, in a sustainable place, special effort is made to create and preserve places, rituals, and events that foster greater attachment to the social fabric of the community. The presence of one local gathering spot, whether it is the town square or a downtown coffee shop, can serve as a powerful focal point for community attachment. This concept [is] what Ray Oldenburg (1989) has termed "the third place."

## SUSTAINABLE PLACES ARE INTEGRATIVE AND HOLISTIC

Sustainable communities employ strategies and solutions that arc integrative and holistic. They seek ways of combining policies, programs, and design solutions to bring about multiple objectives. No longer can we make, for example,

transportation decisions isolated from land use considerations. Instead of endlessly constructing more roads and highways in response to traffic congestion, a sustainable community begins to examine ways that land use and development decisions can reduce such demand in the first place, and that transportation investments and land use can be integrated and coordinated to accomplish a variety of community goals, including air quality enhancement, open space preservation, and affordable housing.

A sustainable community is one that looks beyond narrow, conventional solutions to social and environmental problems and addresses them instead from a broad, holistic viewpoint. The issue of affordable housing, for instance, is typically viewed from a narrow housing perspective—the need to keep down the cost of new construction. But affordable housing and an affordable life are not synonymous. Affordability may be enhanced more dramatically through the development of an urban environment that does not require car ownership and all of its attendant expenses. Furthermore, affordable housing by itself may be counterproductive or even irrelevant without ready access to employment, services, open space, and other elements of a high quality of life. Promoting such activities as urban infill, adaptive reuse, and construction of secondary housing units could simultaneously provide more affordable living units and also reduce development pressures on environmentally sensitive lands.

The fragmented nature of planning policy has tended to discourage such integrated and holistic approaches (e.g., zoning decisions made separately from transportation, taxation, and other public investment decisions). A frequently cited example of a city in the developing world that has sought to implement more integrative policies is Curitiba, Brazil. Each sustainability initiative there addresses several concerns simultaneously. For instance, the city buys trash from the urban poor, providing not money but food or bus tickets in exchange. The city thus deals with solid waste and sanitation cheaply, while serving the food and transportation needs of the poor (Rabinovitch 1992). Curitiba illustrates how important integration, holism, and creativity arc in moving toward sustainability.

## SUSTAINABLE PLACE IMPLIES A NEW ETHICAL POSTURE

Sustainability is fundamentally about adopting a new ethic of living on the planet. This ethic expands substantially the "moral community" to which respect and duties are owed. Almost every definition of sustainability and sustainable development implies a substantial extension of the moral community to include future generations—perhaps many future generations—in the decision making of the present. Sustainability significantly focuses our moral attention away from the present, and from short-term time frames, to much longer time horizons. It is an ethical posture much more respectful of the natural world. Aldo Leopold (1949) foreshadowed much of this thinking some sixty years ago when he talked of the need for a new land ethic, one that would change *Homo sapiens* from conqueror of the land community' to "plain citizen" thereof.

A sustainable community ideally embodies and implements an ethical framework in which physical and social form, consumption of land and resources, and basic operating principles arc severely moderated with the interests of the future in mind. This framework also imagines other ways of expanding the moral community: geographically and spatially, for instance. A sustainable community considers the impacts of its actions and policies on neighboring jurisdictions and the region in which it is situated, as well as its continental and global impacts. Such morally inclusive visions are illustrated in the actions of the Chesapeake Bay communities that consider the impacts their developments will have on the Bay's water quality, or the efforts of cities such as Toronto and Portland, Oregon, to address their role in global climate change by curtailing carbon dioxide emissions. (For a more extensive discussion of expanding the moral community and ethical land use more generally, see Beatley 1989, 1994b.)

In many ways, then, the idea of a sustainable community represents a deeper and more modern vision of Leopold's land ethic. The policy implications of such a new ethical posture are many. Calls for five-hundred-year planning (Tonn 1986) no longer seem ludicrous, but perhaps even conservative. Such a moral time frame is clearly different from traditional comprehensive local planning, which considers the relatively short time intervals of fifteen, twenty, or twenty-five years. Looking so far into the future may also call for new planning techniques and visioning methods. Such a new ethical perspective emphasizes creating urban environments of enduring value, protecting the basic environmental and ecosystem functions essential for existence, and avoiding irreversible actions such as destruction of natural landscapes, species extinction, and the loss of cultural and historic resources. Such a perspective implies caring for soils, forests, oceans, and other life-giving renewable resources; reducing waste at all levels; and restraining ourselves from passing along an increasingly toxic environment to those who follow us.

The ethical underpinnings of sustainability go beyond obligations to the environment. The ethic also emphasizes equity in the distribution of social goods and resources, an effort to improve the lot of the least advantaged in society and eliminate environmental and other forms of racism.

How the new ethic will take hold is uncertain. ... Education, public discourse, and advocacy by the planning field are some of the likely avenues. As fanciful as they sometimes seem, we should consider mechanisms that will include explicit representation of the interests of future generations—a sort of future-generations ombudsperson, something like the Seventh Generation rule of the Iroquois tribe (see Beatley 1994a). Also critical is an ethic of place building and nurturing a responsibility to place and to community. The principles of bioregionalism—which emphasize the importance of identifying with, understanding, and committing to one's bioregional "home"—are helpful in this regard (e.g., Sale 1991).

## Sustainable Places Strive to Be Equitable and Just

Sustainability must also be about creating a more equitable and just society. We must always ask, Sustainable for whom? and be cautious that any vision of a sustainable community be an accessible one: one that is open to all racial, cultural, age, and income groups and that encourages social and cultural diversity. It is a place that strives to be gender neutral and to ensure physical access and social opportunity to all its members. A sustainable community, therefore, is a just and equitable community.

On the whole, contemporary patterns of urban development in the United States raise questions about social justice and fairness. With the march of sprawl and suburbanization has also come sharp separation and isolation of the poor and minorities (e.g., Downs 1994; Goldsmith and Blakely 1992). And there is growing recognition that, while the causes of these trends are complex, many features of traditional planning policy discussed earlier—auto and highway dependency and the separation of

land uses, for example—have played important roles.

A sustainable community, then, is one in which diversity is tolerated and encouraged, where there is no sharp spatial separation or isolation of income and racial groups, where all individuals and groups have access to basic and essential services and facilities, and where residents have equality of opportunity (Beatley 1993).

Issues of equity also arise when considering the current patterns of use and exploitation of the world's resources, and these are not irrelevant in discussions about U.S. growth and development patterns. It has been frequently observed that, while the developed world accounts for only a small percentage of the world's population, its use of natural resources and its generation of waste and pollution is disproportionately high. Industrialized nations use more resources and pollute more than less developed countries in almost all categories. Per capita fossil fuel consumption for industrialized countries is almost ten times what it is for developing nations: use of nonrenewable resources such as cooper and aluminum is some seventeen and twenty times greater, respectively; and consumption of roundwood (wood from trees) is two-and-a-half times greater—although consumption rates for many of these resources are on the rise in the developing world (WRI 1994). Per capita emissions of carbon dioxide are also much higher; the U.S. rate is some nine times that of China and eighteen times that of India (WRI 1994). An oft-cited statistic is that about one-fifth of the earth's population is currently living in "absolute poverty," or with an annual income less than U.S. $370 per year (Carley and Christie 1993). Responsible local and regional efforts at promoting sustainability have the potential to reduce inequities that exist at both global and local levels.

Many believe that we have gone well beyond consuming our fair share, and that movements toward greater sustainability (and the promotion of sustainable communities) are called for on these grounds of equity. At an urban or local level, it is clear that many communities are able to maintain their lifestyles and consumption levels only by "appropriating" the carrying capacities of regions and communities beyond their borders, again raising basic questions of fairness (Roseland 1992; Rees 1992, 1995).

## Sustainable Places Stress the Importance of Community

The concept of community is central to the vision of sustainable communities. Contemporary development patterns do not create communities, although they may create developments. Typical patterns of suburban growth are viewed by many as antithetical to the creation of places where people can share a true connectedness with others and develop feelings of responsibility and closeness. Development and consumption patterns of the last fifty years reflect the long-standing American celebration of individualism. A paradigm of sustainability renews the commitment to community and rises above narrow individualism.

Amitai Etzioni has emerged as a leading spokesperson for the communitarian movement, articulating its key principles and beliefs most clearly in *The Spirit of Community* (1993). A major communitarian premise is that society has moved too far in the direction of individual rights and away from notions of personal responsibility and commitment to a larger "we-ness." As Etzioni asserts, "Americans are all too eager to spell out what they are entitled to but are all too slow to give something back to others and to the community" (1993, p. 15). The time has come, the communitarians believe, "to

attend to our responsibilities to the conditions and elements we all share, to the community" (Etzioni 1993, p. 15; see also Galston 1991 and Etzioni 1988). Moving toward greater community does not, however, mean quashing individualism—only restoring a sensible balance between the desires of the individual and the needs of the community.

The implications of the communitarian ideal are considerable, and Etzioni and others have offered a number of specific changes and initiatives to advance the agenda. At the heart of it would be places where people are committed, invested, and involved, and places where people know and care about one another, participate in community activities, and take responsibility for the condition and health of the community and the environment. The agenda has both personal and public policy dimensions.

Obviously, a community's physical form and design influences the opportunities for true community. Low-density, scattered, cul-de-sac development patterns and a dearth of public spaces encourage isolation and discourage interaction. Sitting in bumper-to-bumper highway traffic may induce a sense of mutual empathy among commuters but does not create a community among them. The physical characteristics of a sustainable community help to create a sense of community—a sense of ownership, commitment, and a feeling of belonging to a larger whole. Walking spaces, civic buildings, plazas and parks, and other public places have the potential to nurture commitment and attachment to the larger collective.

The centrality of community suggests the need to look for a new politics—a politics more consistent with the communitarian spirit. DeWitt John (1994) calls for "civic environmentalism" as a response to both the command-and-control, top-down federal approaches pursued in programs such as the federal Clean Air and Water Acts and the highly personalized and localized politics of NIMBYism ("not in my backyard"). Sustainable communities—as both a vision and a movement—appear to hold considerable promise as an alternative form of politics concerned with the full range of issues that affect quality of life community-wide and a politics that has the potential to involve and be relevant to all or most individuals and groups. Community sustainability may offer a useful unifying framework that embodies a more integrative and holistic viewpoint, in which the health of the larger community is what becomes most important.

Community sustainability as a political movement is still very much in its infancy, but the concept has stimulated a great deal of local activism. Throughout the country, local organizations are emerging that haw the ability to formulate a comprehensive and integrated vision, to transcend NIMBYism, and to be inclusive of a wide range of local groups and interests—from environmental groups to neighborhood groups to housing advocates to civil rights organizations. The number and strength of these groups will likely grow—a positive trend toward a new politics that is inclusive and more reflective of the true values of community.

## SUSTAINABLE PLACES REFLECT AND PROMOTE A FULL-COST ACCOUNTING OF THE SOCIAL AND ENVIRONMENTAL COSTS OF PUBLIC AND PRIVATE DECISIONS

What are the true environmental and social costs of air pollution, water pollution, and destruction of wetlands and wildlife habitat? And, similarly, what are the true social costs of racism, and of separation and isolation of income and social groups, and of a sterile and uninspiring urban landscape? A sustainable community seeks to assess and understand these full costs and to adjust its planning and

other decisions accordingly. It attempts, wherever possible, to promote full-cost accounting in its decisions and policies, using the power of economic signals and mechanisms in promoting greater sustainability.

While modifying the economic and market signals to better account for the true costs of public and private decisions is not the entire answer to promoting sustainability, it will certainly facilitate a movement in that direction. Understanding how the many current subsidies and pricing policies for auto transport influence the choice of this mode over mass transit, for example, may lead to adjustments that support greater sustainability. There is a substantial advocacy for full-cost environmental accounting, and the tools and techniques for doing so have consistently improved (see Daly and Townsend 1993; Kopp and Smith 1993; Cairncross 1992; Pearce, Markandya, and Barbier 1989). Examples include the use of "green fees" to account for the disposal of garbage and household wastes, local development impact fees to shift development in the direction of more sustainable sites and designs, and regional tax base sharing to encourage coordinated land use planning. The following chapters explore these techniques in greater detail.

## CONCLUSIONS

The concept of sustainable places means different things to different people. While the reader may not agree with every element presented in this chapter, there is a range of qualities and factors that should be addressed in planning for sustainable communities. Without some clarity and social consensus about the characteristics of such places, it will be difficult to achieve a more positive result. The principles articulated here suggest a better model for planning and managing in the future, and vast improvement over our current way of thinking about communities. The paradigm expressed here is both social and environmental, seeking a union of goals. The paradigm is also necessarily normative—that is, it explicitly expresses certain values and ethical responsibilities, including duties to live within ecological limits, to consider generations yet to come, to value the equity of our current relationships, and to rise to the demands of community.

## REFERENCES

Baldwin, A. Dwight, Judith DeLuce, and Carl Pletsch, eds. 1993. *Beyond Preservation: Restoring and Inventing Landscapes.* Minneapolis: University of Minnesota Press.

Beatley, Timothy. 1994a. "Creating Sustainable Communities." *Colonnade* X (i) (Spring): 7–13.

Beatley, Timothy. 1994b. *Ethical Land Use: Principles of Policy and Planning.* Baltimore: Johns Hopkins University Press.

Beatley, Timothy. 1993. "Urban Policy and Fair Equality of Opportunity," In *Shaping a National Urban Agenda,* edited by Gene Grigsby and David Godschalk. Los Angeles: University of California, Los Angeles Center for Afro-American Studies.

Beatley, Timothy. 1989. "Environmental Ethics and Planning Theory." *Journal of Planning Literature* 4 (1): 1–32.

Brown, Lester R., and Hal Kane. 1994. *Full House: Reassessing the Earth's Population Carrying Capacity.* New York: Norton.

Cairncross, Frances. 1992. *Costing the Earth.* Cambridge, MA: Harvard Business School Press.

Carley, Michael, and Ian Christie. 1993. *Managing Sustainable Development.* Minneapolis: University of Minnesota Press.

Daly, Herman E., and Kenneth N. Townsend, eds. 1993. *Valuing the Earth: Economics, Ecology and Ethics.* Cambridge, MA: MIT Press.

Downs, Anthony. 1994. *New Visions for Metropolitan America.* Washington, DC: The Brookings Institution.

Etzioni, Arnitai. 1993. *The Spirit of Community: The Reinvention of American Society.* New York: Touchstone.

Etzioni, Arnitai. 1988. *The Moral Dimension: Toward a New Economics.* New York: Free Press.

Galston, William A. 1991. *Liberal Purposes: Goods, Virtues and Diversity in the Liberal State.* New York: Cambridge University Press.

Godschalk, David R., and Francis Parker. 1975. "Carrying Capacity: A Key to Environmental Planning?" *Journal of the Soil and Water Conservation Society* 30(4): 160–165.

Goldsmith, William W., and Edward Blakely. 1992. *Separate Societies: Poverty and Inequality in U.S. Cities.* Philadelphia: Temple University Press.

Jacobs, Allan B., 1993. *Great Streets.* Cambridge, MA: MIT Press.

John, DeWitt. 1994. *Civic Environmentalism: Alternatives to Regulation in States and Communities.* Washington, DC: CQ Press.

Jordan, William. 1993. "'Sunflower Forest': Ecological Restoration as the Basis for a New Environmental Paradigm." In *Beyond Preservation: Restoring and Inventing Landscapes,* edited by Baldwin, Deluce, and Pletch. Minneapolis: University of Minnesota Press.

Kopp, Raymond J., and V. Kerry Smith, eds. 1993. *Valuing Natural Assets: The Economics of Natural Resource Damage Assessment.* Washington, DC: Resources for the Future.

Kunstler, James Howard. 1993. *The Geography of Nowhere.* New York: Touchstone.

Leopold, Aldo. 1949. *A Sand County Almanac.* New York: Oxford Press.

Marsh, William M. 1991. *Landscape Planning: Environmental Applications.* New York: John Wiley.

McHarg, Ian. 1969. *Design with Nature.* Garden City, NY: Anchor.

Oldenburg, Ray. 1989. *The Great Good Place.* New York: Paragon House.

Pearce, David, Anil Markandya, and Edward B. Barbier. 1989. *Blueprint for a Green Economy.* London: Earthscan.

Rabinovitch, Jonas. 1992. "Curitiba: Towards Sustainable Urban Development." *Environment and Urbanization* 4(2): 62–73.

Rees, William E. 1995. "Achieving Sustainability: Reform or Transformation?" *Journal of Planning Literature* 9(4): 343–61.

Rees, William E. 1992. "Ecological Footprints and Appropriated Carrying Capacity: What Urban Economics Leaves Out." *Environment and Urbanization* 4(2): 121–30.

Roseland, Mark. 1992. *Toward Sustainable Communities.* Ottawa, Ontario: National Round Table on the Environment and the Economy.

Sale, Kirkpatrick. 1991. *Dwellers in the Land: The Bioregional Vision.* Philadelphia: New Society.

Salvesen, David. 1990. *Wetlands: Mitigating and Regulating Development Impacts.* Washington, DC: Urban Land Institute.

Schneider, Devon M., David R. Godschalk, and Norman Axler. 1978. *The Carrying Capacity Concept as a Planning Tool.* PAS Report 338. Chicago: American Planning Association.

Spirn, Anne Whister. 1984. *The Granite Garden.* New York: Basic Books.

Steiner, Frederick. 1991. *The Living Landscape: An Ecological Approach to Landscape Planning.* New York: McGraw-Hill.

Thurow, Charles, William Toner, and Duncan Erley. 1975. "Performance Controls for Sensitive Lands." PAS Report 307/308 Chicago: American Planning Association.

Tonn, Bruce E. 1986. "500-year Planning: A Speculative Provocation." *Journal of the American Planning Association* 52(2): 185–93.

World Resources Institute. 1994. *World Resources, 1994–95.* New York: Oxford University Press.

# THE COSTS—OR HAVE THERE BEEN BENEFITS, TOO?— OF SPRAWL

By Alex Krieger

In the growing literature on sprawl, a predominant view holds urban sprawl accountable for much that is wrong with urban America. This is the view of New Urbanists, among others, who consider sprawl a recent and aberrant form of urbanization that threatens even the American Dream. Such is the thesis of the oft-cited *Suburban Nation: The Rise of Sprawl and the Decline of the American Dream* (2000) by Andreas Duany, Elizabeth Plater-Zyberk and Jeff Speck, and of their West Coast counterparts Peter Calthorpe and William Fulton in *The Regional City: Planning for the End of Sprawl* (2001).[1]

A second view—today less often expressed by planners or the media—is that the effort to control sprawl is an elitist attack on the American Dream, an attack that withholds that dream from those who are still trying to fulfill it. Its current spokesmen are libertarians and others opposed to further government restrictions on property rights.

While opposition to sprawl is growing, the motivations for this opposition are complex and occasionally contradictory. And while support for, or acquiescence to, sprawl generally comes from those fighting to maintain unencumbered property rights, their reassertion of the benefits of sprawl—benefits that motivated most American land development in the first place—cannot be so easily dismissed.

But what constitutes sprawl? That simple word carries the burden of representing the highly complex set of effects from low-density urban expansion. Humanity is still urbanizing with cities worldwide spreading outward at unprecedented rates, but in North America, sprawl, though not literally synonymous with suburbanization, generally refers to suburban-style, auto-dominated, zoned-by-use development spread thinly over a large territory, especially in an "untidy" or "irregular" way.[2] Among the oldest and most persistent critiques of American urban sprawl centers on this visual awkwardness and conjures up an image of the human body *sprawling*.

Mainstream media attention to sprawl, more than citizen attention, has increased dramatically in recent years. Indeed, in the two-years that straddled the Millennium, sprawl was the subject of lengthy articles in such publications as the *Atlantic Monthly*, *Harper's*, *National Geographic*, *Scientific American*, and *Time*, and of several front-page stories in *USA Today*—an impressive attention to land-use by media that generally ignore the subject. Scores of other "something-must-be-done-about-sprawl" features, including two Ted Koppel *Nightline* shows, appeared during the period. *Preservation Magazine* even chimed in with a long essay on "Golf Sprawl."[3] On the heels of (and perhaps because of) a decade of prosperity, and as Americans faced a new century, the media identified sprawl as that condition of urbanization that was producing— and if allowed to continue would rapidly accelerate—an erosion in Americans' quality of life. A seductive sound bite to counter sprawl also continued to gain prominence: "Smart Growth." Around this mantra gather environmentalists, proponents of urban reinvestment, advocates of social equity, preservationists, spokesmen for various "livability agendas,"[4] public housing officials, and a few trend-sensitive developers, all rallying against, well, sprawl. At the turn of the millennium, those who consider themselves enlightened about land use and environmental stewardship view sprawl as bad for America.

More recent concerns about security and a weaker economy have shifted public and media attention, but have not relegated discussion of sprawl back to planning journals. Indeed, one of the worries among city advocates immediately following the events of September 11, 2001, was a potential re- acceleration of suburbaniza-tion—of people and businesses seeking "safer" places to live and work than terrorist-target areas like Manhattan. This reaction, perhaps over-reaction, has a precedent during the Cold War when the threat of nuclear holocaust produced similar concerns, launching campaigns for "defensive dispersion." In the late 1940s and early 1950s, planning journals (and scientific journals like *Bulletin of the Atomic Scientists*) regularly published articles like "The Dispersal of Cities as a Defensive Measure" and "A Program for Urban Dispersal."[5]

I will return to the arguments periodically made on behalf of sprawl. But what are the arguments against it? There are five principal lines of critique: The first and oldest is aesthetic, though not often recognized as such. Recall the "ticky-tacky houses" folk songs of the 1960s and, earlier still, the damning words of a poet, relevant still three-quarters of a century later: "I think I shall never see/A billboard lovely as a tree/Perhaps, unless the billboards fall,/I'll never see a tree at all."[6] While there is a trace of eco-logical concern in these lines by Ogden Nash for many, even those lacking poetic sensibilities, the physical environments produced by miles of low-density settlement are simply ugly. They disfigure and insult both nature and worthier examples of human artifice. Among the most effective tactics used by New Urbanists is to simply produce images of prettier environ-ments—recalling the charms but never the limitations of old small towns. Such Currier & Ives vignettes of the future (rather than, as the originals portrayed, of scenes of rapidly disap-pearing vernacular traditions) help persuade some that the character of places vanished can be recovered to replace the visual chaos of the contemporary suburban landscape. Whether the dressing up of the suburb in town-like ico-nography can actually diminish sprawl remains to be proven. It seems unlikely that more attractive or even more compact, subdivisions would significantly reduce Americans' appetite for roaming far and wide in search of either necessities or amusements.

The second argument is sociological. Already in the 1950s, critics like William H. Whyte and John Keats portrayed suburban life as conformist, drab and isolationist.[7] In the decades since, such arguments have expanded to suggest correlation between suburbanization and social apathy, intolerance of neighbors unlike oneself, segregation, and so forth. Concerns are voiced about alienated suburban youth, dependent on parent chauffeurs to get anywhere, about the enslavement of parents to their chauffeur role, and about the isolation of grandparents who can no longer drive themselves. Apprehension about the social isolation of suburban stay-at-home moms has gradually shifted to sociologists' worry about the difficulties of combining careers and child rearing across a dispersed landscape. The title of Robert Putnam's recently popular *Bowling Alone* implies that privation of group activity is also a consequence of lives spent in sprawled, disconnected America, although Putnam could draw only circumstantial correlation between sprawl and a decline in civic engagement.[8]

The third critique is environmental and remains the most compelling. This critique has slowly (far too slowly for some) gained power since the late '60s and early '70s when Rachel Carson's *Silent Spring*, Ian McHarg's *Design With Nature*, the first Earth Day, and publications such as *The Limits to Growth* and *The Costs of Sprawl* helped arouse profound concern about human abuses of the environment.[9] Although worldwide environmental degradation has many causes, sprawl is certainly a contributor. Few can argue that low-density development does not increase auto emissions, water use, pollution, trash, loss of species habitat, and energy consumption. To cite one example, most pollution of ground water, lakes, streams, and rivers in the United States is caused by runoff that collects various toxins on the high percentage of impervious surfaces, like roads and

parking lots, in urbanized regions. The heating and cooling of freestanding homes, with their many exterior walls per capita, requires more energy than attached, denser development. And then there are those immaculate lawns that require ample water and chemicals to maintain. Of course, most such conditions are caused by increasing affluence, not just settlement patterns, though affluence and sprawl are not unrelated. Environmentalists have become among the fiercest critics of sprawl, armed with sobering statistics and demanding reform. The first year-2000 issue of *Sierra*, the magazine of the Sierra Club, devoted itself entirely to the arrival of what it called "The Green Millennium," which various authors said needed to be freer of sprawl.[10] There is little doubt that calls for better environmental stewardship—leading to legislated restrictions on development—will increase in the coming decades, influencing urbanization patterns considerably.

The fourth argument is that sprawl leads to boring "lifestyles." In addition to dyed-in-the-wool urbanists (like me), some among the generation of now grown children of Baby Boomers, having being raised in the suburbs, are pining for more convivial surroundings. Precisely what proportion feel this way is hard to establish, but various informal housing preference surveys along with the modest recent rise in demand for downtown housing provides considerable anecdotal evidence. A century ago rural populations were lured to cities mainly by economic opportunity. Now younger adults, less inclined to follow in the footsteps of their suburbia-pioneering parents, seek out the cultural and social stimulation of city life. Think of the sultry allure of New York in the TV series "Sex and the City." By comparison, where is the action along Boston's "Technology highway" Route 128 once the day's work of inventing technologies or investing venture capital is done? Rarely does one find fine dancing or music clubs among the

Blockbusters, Burger Kings, and karaoke bars of suburbia. Exoticism is associated with city life among young Americans, even though child rearing years and the accompanying search for better public schools and housing affordability return most to the comforts of suburbia. Though back in the suburbs, young parents maintain their desire for more interesting lives, lamenting how hard these are to assemble amid the sprawl.

The fifth case against sprawl, becoming more prevalent, is self-protection. Outwardly it is waged as a campaign, mostly in affluent communities, against loss of open space and growing traffic congestion. Its underlying stance is less noble, constituting some variation of "don't harm my lifestyle by replicating the locational decisions I made a few years earlier; your arrival will ruin my lovely neighborhood." As David Brooks, the author of *Bobos in Paradise*, noted in a recent *New York Times* article about exurban voters, "Even though they often just moved to these places, exurbanites are pretty shameless about trying to prevent more people from coming after them."[11] On one level, this is understandable. No one wants one's access to nature obstructed, or a commute to work lengthened. However, such a "Not In My Back Yard" attitude pushes development away from areas resisting growth, increasing rather than containing sprawl. New subdivisions simply leapfrog to the next exit along the highway, where less expensive land (along with fewer constraints on development) is available. Once settled these newcomers will guard against subsequent encroachers.

While anti-sprawl literature relying on one or more of these positions receives substantial attention, little fanfare accompanied the recent publication of a rare rebuttal. In 2001, Randal O'Toole, expressing views that have traditionally been mainstream—and may, indeed, still be—published *The Vanishing Automobile and Other Urban Myths*, subtitled *How Smart Growth Will Harm American Cities*.[12] The book's sensibility seems out of kilter with the times, yet its copiously assembled statistics are impressive, if hard to corroborate.

The book calls many of the core assumptions advanced by the critics of sprawl myths. For example, while Jane Holtz Kay's *Asphalt Nation*, a characteristic condemnation of sprawl, cites numerous (equally hard to substantiate) statistics about what she calls "the cost of the car culture," O'Toole asserts that, on a passenger per mile basis, public dollars in support of transit are double what they are for highways.[13] Determining in precise monetary terms how much our culture subsidizes auto usage is nearly impossible. We certainly *favor* car usage, and thus, no doubt, support and benefit from some direct and many indirect subsidies. Still, within the narrow terms of how he frames the issue— passenger per mile costs—O'Toole's makes his point clear: Since most of us use cars and few of us use public transit, the *public* investment in public transit *per user* is plausibly higher than the public investment *per user* for highways. This doesn't mean (although O'Toole would so argue) that it is not sound public policy to invest in public transit or raise the cost of driving.

Public subsidy of auto usage is but one of the seventy-three (!) myths that O'Toole identifies in what he calls the "The War Against the Suburbs."[14] A few other examples: He criticizes the much admired experiment in regional growth management in Portland, Oregon, by pointing out (as others have) the resulting rise in housing costs in the center of the city and the fact that light rail system extensions have reduced the number of neighborhood bus lines. He concludes that both changes disproportionately affect the poor, and he thus questions the social equity arguments advanced by transit proponents and growth boundary advocates. He quantifies the substantial preference that

Americans at almost all social and economic levels continue to show for larger homes, less density, more open space, and the personal wealth generation that home-ownership has brought. He debunks the assertion that new highways increase congestion by attracting additional traffic (first claimed by Lewis Mumford in the 1950's[15]) by pointing out that over the past two decades, while the number of auto miles traveled has nearly doubled, the number of road miles has increased by less than three percent. He has the temerity to suggest that people *like* to drive, rather than being *forced* to drive by an absence of alternatives. He points out that less than five percent of the land area of the continental United States is urbanized, so fears of running out of land are premature. He argues that it is density, not dispersion, that causes congestion, offering statistics that the densest American cities have the worst incidence of congestion and often the longest commutes. In a characteristic dig at conventional smart growth wisdom, which supports density and opposes highways, he writes: "The Los Angeles metropolitan area [must be] the epitome of smart growth, as it has the highest density and the fewest miles of freeway per capita of any U.S. urbanized area."[16]

To anyone whose values or intuitions align with current critiques against sprawl, O'Toole's conclusions seem either irresponsible or naively contrarian. Of course, the arguments for and against sprawl are not going to be resolved by competing value-laden statistics. As the furious debate fueled by the publication of Bjorn Lomborg's *The Skeptical Environmentalist* illustrates, ideology and polemical bias can bend many a statistic.[17] Dismissing O'Toole's stance, nonetheless, disregards that for much of American history, sprawl (though not called that) was considered progressive, a social good, and a measure of citizens' economic advancement.

Prior to the concern about population concentrations brought about by the atomic bomb, during the 1930's for example, President Roosevelt's Resettlement Administration was committed to sprawl—then called *decentralization*. It was seen as one means for recovering from the Great Depression and preventing similar economic setbacks in the future.[18] A widely held assumption was that among the causes of the Depression were unwieldy and unmanageable concentrations of commerce, capital, and power. In other words, many concluded that huge unmanageable cities (like New York) were partially to blame.

Two generations earlier, Henry George, writing in *Progress and Poverty,* predicted that concentration of urban populations would worsen economic inequality. He argued passionately that social inequality was endemic to cities, where overcrowding and land possession by the few perpetuated poverty. His "remedies" were to eliminate all private land ownership (impractical, of course) and to disperse urban populations, so that "The people of the city would thus get more of the pure air and sunshine of the country, and the people of the country more of the economic and social life of the city."[19] For George's many followers, and the American advocates of the slightly later Garden City Movement, the road away from inequality led out of cities. This argument echoes even today in the continuing migration from older urban centers of people in search of economic upward mobility.

The affirmation of population decentralization can be traced in a straight intellectual line to America's founding fathers, in particular to the persuasive Thomas Jefferson. Fearing the consequences of America becoming urban, Jefferson went so far as to invent a land-partitioning policy that he hoped would negate the need for urban concentrations. For Jefferson, cities were corrupting, even "pestilential," influences and government support for the small landowner—dispersed on his self-sufficient homestead—was crucial to America's future.[20]

Jefferson's worries about urbanization seemed prescient to those witnessing the unprecedented urban concentrations of the later part of the 19th and early part of the 20th centuries. At the turn of the 20th century, daily life in New York's Lower East Side, depicted in Jacob Riis's photojournalism and like the life in London's slums depicted in Charles Dickens' novels, offered little hope for improving the human condition. What seemed problematic about contemporary urbanization prior to the mid-20th century (and what remains problematic in much of the developing world) was concentration. And sprawl, although called by various less tarnished names, was advocated as a partial solution.

Thus, by the time the middle class sprawled outside cities in great numbers in the decades following World War II, widespread public optimism about the results prevailed, despite an occasional dissent from a William H. Whyte or a Lewis Mumford. It is eerie now to read Whyte's 1958 (!) essay in *Fortune*, entitled "*Urban Sprawl*," or John Keats' 1957 novel *A Crack in the Picture Window*, or Peter Blake's 1963 "*The Suburbs are a Mess*" in *The Saturday Evening Post*.[21] Much of the aesthetic and social arguments against sprawl (the ecological perspective arose about a decade later) were already well enumerated, or at least anticipated, a half-century ago. Very few citizens were paying much attention, however. Quite happily, and by the millions, Americans sought out the comforts, spatial expanse, clean air, economic leverage, and *novelty* of the Levittowns and their various imitations. At mid-20th century, sprawl was considered good for Americans and the nation.

What then has made that optimism (a fulfillment of the Jeffersonian ideal) wane, and has it truly waned?[22] Have Americans actually adjusted their image of the good life and its setting? Are contemporary critics of sprawl that much more eloquent than Lewis Mumford or William Whyte? Hardly. What has changed is the impact on individuals caused by the sprawl of others. What has changed is the quantity of sprawlers and the sheer scale of their sprawling.

In the half-century since 1950, the spread of sprawl has been exponential. Urban populations slightly more than doubled, while the land area used by this population has increased by a factor of four! In the Los Angeles area the factor was seven! Two million acres of farmland and open space have, and are continuing to be, lost to development every year. Cars have multiplied twice as fast as the population. Estimates of the costs of time lost and fuel wasted in traffic range into the billions of dollars per year. Ozone-alert days in sprawled metro areas such as Atlanta or Phoenix have been rising for decades, despite improved auto emissions and other environmental controls. North Americans currently use the equivalent of ten acres of land per capita, whereas less developed countries use approximately one acre per capita.[23]

Such disturbing statistics have only recently countered the complacency of suburbanites, or wannabe ones, who heretofore believed that by simply moving further out they could avoid the personal inconveniences caused by the sprawl. For most Americans it has always been easier to retreat than to repair. This has lead to schizophrenic urbanism—people making new places that evoke old qualities while being oblivious to the consequences of abandoning exemplary places made earlier. This self-perpetuating cycle of American urbanization—disinvestments in settled areas, expanding rings of new development, wasteful consumption of resources, obsolescence, highway congestion, economic (now more than racial) segregation, homogeneity, ugliness, all leading to new cycles of perimeter development-is finally being acknowledged by more of us as self-defeating.

What has begun to rattle Americans is the awareness that once everyone got "out there," some of the advantages of "getting away" have proven illusive. This, however, does not mean that Americans believe that such advantages are no longer worth pursuing, as Randal O'Toole or *USA Today* remind us. In a recent *USA Today* survey giving people four choices of ideal living circumstances, fifty-one percent chose a 100 year-old farm on ten acres, thirty percent chose a five-bedroom Tudor in the suburbs, 10.5 percent selected a Beverly Hills mansion, and a mere 8.5 percent chose a designer loft in Manhattan.[24] No, Americans have not yet abandoned their sprawling instinct, but they are developing a lower tolerance for the sprawl of their neighbors. This is generally unacknowledged in the waves of anti-sprawl literature (which my wife has labeled "the scrawl about sprawl").

What must be brought to the fore in the debate over sprawl is this: The benefits of sprawl—for example, more housing for less cost with higher eventual appreciation—still tend to accrue to Americans individually, while sprawl's cost in infrastructure building, energy generation, pollution mitigation, tends to be borne by society overall. Understanding this imbalance is essential, and seeking ways to adjust to whom and how the costs and benefits of sprawl accrue remains the real challenge. Can political will be developed on behalf of impact fees, user assessments, regional tax-sharing, higher gasoline taxes and highway tolls, streamlined permitting and up-zoning in already developed areas, ceilings on mortgage deductions, surcharges on second homes, open space (and related) amenity assessments, regional transfer-of-development rights, and similar ideas that may shift some of the costs of sprawl onto the sprawlers? There is infrequent evidence

of this today but there is hope that growing awareness of sprawl will lead to such policies. Yes, continuing to find new arguments against sprawl is valuable, but the campaign to create a more diverse, rewarding, and environmentally sound urban future will ultimately depend on Americans finding ways to calibrate short-term self-interest with long-term social value.

## NOTES

1. Andres Duany, Elizabeth Plater-Zyberk, and Jeff Speck, *Suburban Nation: The Rise of Sprawl and the Decline of the American Dream* (New York: North Point Press, 2000). Peter Calthorpe and William Fulton, *The Regional City: Planning for the End of Sprawl* (Washington, DC: Island Press, 2001).

2. *The New Oxford American Dictionary* definition of *sprawl* emphasize such ungainly, irregular, awkward conditions.

3. Bruce Katz and Jennifer Bradley, "Divided We Sprawl," *Atlantic Monthly*, December 1999, 26–42; John G. Mitchell, "Urban Sprawl," *National Geographic*, July 2001, 48–71; Donald D. T. Chen, "The Science of Smart Growth," *Scientific American*, December 2000, 84–91; Richard Lacayo, "The Brawl Over Spraw," *Time*, March 22, 1999, 44–48; James Morgan, "Golf Sprawl," *Preservation Magazine*, May/June 2001, p. 38–47, 115.

4. In 1998 during the early phases of Vice President Al Gore's presidential bid he published a policy document called "*Clinton-Gore Livability Agenda: Building Livable Communities for the 21st Century*." For various reasons (including the supposition that it did not catch on with the electorate) the smart growth part of his campaign became less and less pronounced over the course of the campaign.

5. Tracy B. Augur, "The Dispersal of Cities as a Defensive Measure," *Journal of the American Institute of Planners*, Summer, 1948, p. 29–35; Donald and Astrid Monson, "A Program for Dispersal," *Bulletin of the Atomic Scientists* 7, 1951, p. 244–250. A recent essay has carefully reviewed such dispersal strategies from mid-century. See Michael Quinn Dudley, "Sprawl as Strategy: City Planners Face the Bomb," *Journal of Planning Education and Research 21*, 2001, p. 52–63.

6. Ogden Nash, *The Pocket Book of Ogden Nash* (New York, Little Brown, 1962). By coincidence, 1962 also was the year Malvina Reynolds' famous folk song, "Little Boxes," "They're all made out of ticky-tacky, and they all look just the same."

7. William H. Whyte, "Urban Sprawl," *Fortune*, January, 1958, 102–109. The terms *urban sprawl* may have been coined with this essay. John Keats, *The Crack in the Picture Window* (New York: Houghton-Mifflin, 1957). Keats' novel so railed against the disfunctionalities of suburban lifestyles that he compared suburbia to the urban nightmare in George Orwell's *1984*.

8. Robert D. Putnam, *Bowling Alone: The Collapse and Revival of American Community* (New York: Simon & Schuster, 2000). Putnam postulated several causes for an increase in civic disengagement, but concluded: "Yet [sprawl] cannot account for more than a small fraction of the decline, for civic disengagement is perfectly visible in smaller towns and rural areas as yet untouched by sprawl" (215).

9. The first Earth Day was held in 1970. Rachel Carson, *Silent Spring* (New York: Fawcett Crest, 1962).Ian McHarg, *Design With Nature* (New York: National History Press, 1969). Donella H. Meadows, Denis L. Meadome, Jorgen Randers, William W. Behrens III, *The Limits to Growth: A Report for the Club of Rome' Project on the Predicament of Mankind* (New York: Signet, 1972). Real Estate Research Corporation.

*The Costs of Sprawl in Detailed Cost Analysis* (Washington D.C.: Government Printing Office, 1974).

10. Curbing sprawl was one of "five bold ideas for the new century" offered in the January/February 2000 issue of *Sierra*.

11. David Brooks, "For Democrats, Time to Meet the Exurban Voter," *The New York Times*, November 10, 2002, 3.

12. Randal O'Toole, *The Vanishing Automobile and Other Urban Myths: How Smart Growth Will harm America* (Bandon, OR: Thoreau Institute, 2001)

13. Jane Holtz Kay, Asphalt Nation: How the Automobile Took Over America and How we can take it Back (New York: Crown Publishers, 1997). O'Toole, p. 117.

14. O'Toole, 37.

15. Lewis Mumford first addressed the car as a "destroyer of cities" in a 1945 publication entitled *City Development*, and expanded the argument in an article entitled "the Highway and the City" published in *Architectural Record*, April 1958. He devoted considerable attention to the issue in his classic *The City in History: Its Origins, Its transformation, and its Prospects* (New York: Haracourt brace Jovanovich Publishers, 1961), and most vehemently in The Highway and the City (Westport, CT: Greenwood Press, 1981, first published 1963).

16. O'Toole, 392.

17. Bjon Lomborg, *The Skeptical Environmentalist: Measuring the Real State of the World* (Cambridge and New York: Cambridge University Press, 2001) and *Scientific American,* "Misleading Math about the Earth: Science defends itself against *The Skeptical Environmentalist,*" January 2002. Lomborg's book, the lengthy critiques published by 4 scientists in Scientific American, and Lomborg's rebuttal to these critiques unleashed a virtual firestorm of other rebuttals, and an occasional essay in Lomborg's defense, in scores of popular and scientific environmental journals,

and across the Internet that continues to this day.

18. An often repeated statement attributed to Rexford G. Tugwell, President Roosevelt's first administrator of the Resettlement Administration spoke directly to the hopes for decentralization: "… to go just outside centers of population, pick up cheap land, build a whole new community and entice people into it. Then go back into the cities and tear down whole slums and make parks of them." Quoted in Arthur Schlesinger, Jr., *The Age of Roosevelt: The Coming of the New Deal* (Boston: 1958) p. 370.

19. Henry George, *Progress and Poverty: An Inquiry into the Cause of Industrial Depressions and of Increase of Want with Increase in Wealth: A Remedy*, first published in 1879 (New York: Robert Schalkenbach Foundation, 1981). p. 127.

20. Thomas Jefferson often expressed his concerns about a future urbanized America. A typical example is found in a letter to James Madison written in 1787: "I think our government will remain virtuous for many centuries; … as long as there shall be vacant land in any part of America. When they get piled upon one another in large cities, as in Europe, they will become corrupt as in Europe." Quoted in A. Whitney Griswald, "The Agrarian Democracy of Thomas Jefferson", *The American Political Science Reader*, v. XL #4, August 1946, p. 668.

21 W. H. Whyte and John Keats, op cit. Peter Blake, "The Suburbs are a Mess'"The Saturday Evening Post, October 5, 1963.

22. One would think that optimism has waned when reading a report such as the Bank of California (along with several environmental and housing advocacy organizations) sponsored "Beyond Sprawl: New Patterns of Growth to fit the New California" first published in 1995 and widely distributed since then. The executive summary begins with the following sentence: "Ironically, unchecked sprawl has shifted from an engine of California's growth to a force that now threatens to inhibit growth and degrade the quality of life."

23. Statistics gathered from the 2000 U. S. Census: http://quickfacts.census.gov/qfd/

24. *USA Today*, USA Today Snapshots, "Country Calls: Where Americans say they would live if money or circumstances were not an issue", 27 August 2002, p. 1. Survey conducted by Leslie A. Rifkin and Associates.

# GREEN CULTURE
# AND THE EVOLUTION
# OF ARCHITECTURE

By Peter Buchanan

The most overwhelmingly urgent crisis facing mankind is the degradation of the natural environment, including the atmosphere and seas, and the concomitant problems of global warming and climate change. Virtually all other serious problems (such as over-population, hunger, social breakdown and inequality, the rise of diseases such as cancer and the spread of others, increasingly frequent and devastating 'natural' disasters such as storms and flooding) are part of this larger crisis, or closely related and subordinate to it. The continuation of what we would consider to be a comfortable and civilized form of human life depends on us, as individuals as well as collectively as mankind, changing our ways rapidly and radically so as to have a much more gentle impact upon planet Earth by cooperating rather than competing with its natural processes.

Contemporary buildings, like contemporary forms of urban development, are major contributors to the environmental crisis. Resolving this crisis will necessitate transforming all aspects of human settlement, including the construction of new buildings and the retrofitting of old ones, to be as green as possible. It is unthinkable that we can continue to plead that green buildings are unaffordable. Instead we must realize, and quickly that we can no longer afford not to build green.

The green agenda will have immense appeal for architects, once they cease to resist it, because it regrounds architecture in real, immensely serious and urgent issues, and reconnects it with popular desires. These factors, along with the stimulus to fresh thought and creativity they bring, promise to reinvigorate architecture after an extended period of confusion, when many architects were lost pursuing the spurious dictates of fashion and theory. For everybody, green buildings also herald a much-enhanced quality of life, despite being devised to limit wasteful consumption.

Together with being consistent with changes in human values, this is perhaps the most compelling reason for green buildings.

Every week brings fresh evidence of the mounting severity of the global environmental crisis. It has already killed tens of thousands of people, and reduced millions more to misery as refugees from human-provoked 'natural' disasters. The problems confronting us form an all-too familiar and ever-expanding litany of impending disaster. Global warming is redistributing climate systems and so annihilating ecosystems and their life forms, as well as bringing violent storms and summer brush fires, desertification, famine and the spread of killer tropical diseases, along with rising sea levels and the eventual inundation of many of the most densely populated parts of the globe and its greatest cities. Planet-wide pollution and toxic contamination of land, air and water are exterminating more species, and damaging immune systems and chromosomes leading to disease, deformities and infertility in humans and other creatures. Holes in the ozone layer cause cataracts and skin cancer, and threaten to blind insects, thus interrupting the pollination process on which all food cycles depend.

The list could go on and on. Of all the problems we face, these are only some of those to which buildings make an immense and direct contribution. But let us not extend or linger on this litany any further; the anger or debilitating sense of impotence that pondering it can provoke might distract from the main emphasis here, which is on positive and practical action through design.

Facing up to the scale and seriousness of these problems, it is clear that the central challenge of our times is to achieve sustainability. This quest will be a major factor setting the basic tone of the immediate future, and our success or failure in it will decide whether current generations will be admired or reviled in the more distant future. It is a challenge to all of us, individually and collectively: to our imaginations to conceive ways in which sustainability might be achieved, and to our political and corporate will to bring it about.

Yet what is really meant by sustainability? This is among the most abused of current buzzwords. There are architects and manufacturers whose designs and products achieve only small and narrow improvements in environmental performance, and yet are claimed to be sustainable. And there are politicians and corporations whose claims to be acting sustainably are based on the most nebulous, irrelevant and often cynical of reasons that achieve little beyond seductive advertising. Such all-too-common hypocrisies are very damaging, provoking complacency or skepticism, and so a tendency not to take seriously the quest for sustainability. Equally misleading and destructive is the charge that the quest for sustainability is essentially Luddite, and would involve a return to past ways of doing things. Certainly it would involve a new respect for and learning from the past, especially from ways of life once condemned as primitive and now seen as ecologically sound, socially stable and spiritually uplifting in their reverence for the earth and nature. Yet, in part because the planet must now support such a huge human population, sustainability can only be achieved by also moving forward, not just to a better quality of life in accord with emergent values, but often to using leading-edge technologies that are much more efficient than the 'dinosaur' industrial-era technologies that are still so prevalent in construction, manufacture and agriculture.

Basically, sustainability implies long-term viability. As the term sustainable is used today, sustainable developments, cultures, lifestyles or whatever, are those that do not overtax the resources and regenerative capacities of the earth, thus leaving for future generations as much of nature's bounty and beauty as we now enjoy. Achieving such sustainability will require profound changes. For a start, industry, agriculture and city life could no longer depend on extracting and consuming resources, nor on releasing wastes (especially those that are toxic and/or destabilize natural conditions), at rates faster than they can be replenished, or neutralized and absorbed by nature and the earth. Instead we need to learn again to recognize and respect the earth's regenerative cycles and live only within their capacities. In fact, in the medium term at least, we need to allow for more than mere replenishment of what has just been harvested, and give nature a chance to repair the ravages wrought by our overly extractive and polluting economies and lifestyles.

Yet sustainability implies even more than this. Besides living within such natural limits, a sustainable society or culture (which might still be highly dynamic), cannot be vulnerable to excessive instability, social breakdown or a return to ways disrespectful of the earth—all of which would, by definition, render it unsustainable. Thus sustainability depends also on ensuring economic opportunity and social equity, as well as offering everyone a lifestyle that offers more than mere contentment, but sense of meaning and deep satisfaction. Instead of the inequalities and alienation of the present, people would be assured of the means to both fulfill their individual potential and gain a sense of connectedness, of intimate engagement with community and nature and so with their deeper psychological and spiritual selves. These economic, sociological, psychological, cultural and spiritual dimensions are all crucial to sustainability.

There can be no single, exclusive route to sustainability. Just as in nature biodiversity ensures the vitality and adaptability to cope with change and disruptive incidents, so sustainability cannot be achieved by the homogenizing and universalizing tendencies of the waning industrial era. Instead sustainability requires the continued vitality of the earth's various cultures and lifestyles which, although undergoing change and becoming ever more tightly interlinked and interdependent, should remain as diverse as the lands, climates and local traditions each arises from. What is crucial is both to learn from nature (thus conserving biodiversity and emulating it in the human realm) and to see the earth's needs and endless differentiations as primary, and humans as integral to these, arising from and supported by the earth as the conscious component of the flowering forth of evolution.

Mankind's socio-cultural evolution can no longer be driven blindly by the imperatives of the market, consumerism and technological development. Instead we must use the immense accumulated knowledge of our many sciences (including the human sciences such as psychology and anthropology, as well as all the physical and natural sciences) to be more self-consciously aware of and responsible for our own evolution, and so also that of the earth. Just as the great challenge of our times is to achieve sustainability, so that for all responsible and creative people is to participate in this process. Curtailing the profligate use of resources is being achieved by invention as much as by constraining diktat. Participating in this great adventure involves designing and inventing everything afresh: from new social rituals and institutions to such socio-cultural software as new management, economic, legal and tax systems; from new products and manufacturing processes to the transformation of most aspects of the built environment, including building components and transport systems. All of these are essential to, and will help bring about, the various sustainable cultures that will make up a sustainable global civilization. In turn, all of this involves redefining what it is to be human: no longer isolated and alienated from a subordinated and exploited nature, but ennobled and spiritually enhanced by the number and intimacy of our connections with nature. Such are the hugely onerous tasks, and immensely exciting challenges, the environmental crisis bequeaths to us all. These are potentially great gifts to artists and designers of all sorts—be they writers, painters, filmmakers, composers, choreographers, architects, urban or product designers, landscape architects or whatever—who might contribute in various ways to such cultural transformation, thus returning to these creative disciplines a dignity and purpose that has too often been missing of late. If we do manage to tackle the environmental crisis, then we are at the threshold of a creative renaissance.

Clearly, architecture alone cannot bring about sustainability. Nor, for the same reasons that the quest for sustainability leads away from (and not towards) socio-cultural homogeneity, can there be a single, distinct green architecture or green aesthetic. (It is probably safer to talk about green buildings rather than green architecture or, perhaps worse still, sustainable architecture.) Nevertheless, because buildings inevitably impact upon and transform the environment, and because they both express and help shape the lifestyles and values of our culture, buildings and their architects have an immense role to play in the pursuit of sustainability. Yet many architects seem unaware of, or reluctant to acknowledge, how much, and how unnecessarily, buildings contribute to the environmental crisis.

All architects know, of course, that new buildings consume vast amounts of land (all too often, prime arable land needed to feed growing populations), especially when part of the near-ubiquitous sprawl that also necessitates paving vast areas for roads and parking for the gas-guzzling, polluting automobiles it depends upon. Architects also understand that

all this could be done far more wisely than now. They should also recognize that buildings consume and contaminate vast amounts of what is increasingly the most precious of resources, fresh water. They are also vaguely aware that buildings contain many materials that are toxic to manufacture or use (off-gassing noxious fumes—and even worse ones in the eventuality of fire—and polluting rainwater and soil) and that insulating materials and the refrigerants in air-conditioning units were prime culprits in destroying the ozone layer. But too few architects, particularly in the United States, seem at all aware that the construction and operation of buildings is responsible nearly half the energy consumed by developed countries. Moreover, they seem untroubled by an awareness that this is largely unnecessary, as proved by the fact that elsewhere in the world it has become relatively commonplace for a building's energy consumption to be only a small fraction of that of its U.S. equivalent. Some European buildings even harvest the ambient energies of sun and wind so effectively as to be net energy exporters, sending the energy that is excess to their needs to nearby buildings or into the national electricity grid.

Contemporary architecture contributes to the environmental crisis in further, less direct ways than those listed above. Deep-plan air-conditioned buildings, hermetically sealed with tinted glass skins, suppress any sort of sensual contact between their occupants and the natural world outside. Such buildings also tend to be isolated from each other physically and aesthetically, and suppress any form of community life, both—within them and in their surroundings. In alienating people from nature, each other and place, they alienate us also from such essential aspects of human nature as a sense of rootedness and connection. Thus these commonplace buildings, which are almost the contemporary American vernacular, not only mirror but also help entrench the mind-set that has perpetuated, or at least tolerated, the destruction of the natural world. Hence, the design of green buildings (once again, as with achieving sustainability in general) must involve more than resolving technical and ecological issues to also address social and spiritual ones.

Despite the urgency of the environmental crisis, and architecture's vast and varied contribution to both it and its eventual solution, most design professionals have been desperately slow to acknowledge, let-alone take up, these challenges. There are several reasons for this resistance. In the last couple of decades the sort of architecture that has gained attention in the media and academy is that driven by form, fashion and theory (the latter itself largely the subject of fashion as theorists recycle the same limited set of ideas and sources) and has largely turned away from pressing realities, such as social and green issues. Indeed, the latter are stigmatized as untrendy, regressive and even reactionary. (Some academics have gone so far as to refer to ecofascism.) A stereotyped notion of green buildings conjures up images of muesli-eating inhabitants with beards and sandals, and rudimentary forms of back-to-nature lifestyles—a caricature of the counterculture that pioneered much green experiment—as well as of crude and ugly buildings with which no urbane sophisticate or academic would wish to be associated. Although there is more than a grain of truth in such characterizations, they also show ignorance of developments in the rest of this world, such as those that are the focus of this book.

Fears that green design compromises the architect's creative prerogative, and charges that it is also essentially regressive, are commonly encountered. These reflect deep misunderstandings underlying much that is wrong with contemporary architecture. For too many today, creativity—supposedly expressed also in the architect's theories and formal predilections—is exercised as part of the architect's right to self-expression. However, another view of creativity has emerged, especially from the new sciences such as chaos and complexity theories. Creativity is not the prerogative of humans only; instead the whole unfolding of cosmological and natural evolution are seen as essentially creative processes, not just in the long term but moment by moment. Thus the truest and highest form of human creativity is to be found in participating in evolutionary process: transcending the ego in favor of the eco and playing a part in this constant flowering forth, and guiding it to the best advantage of the planet and all its creatures and peoples. Creativity of this sort is less the expression of individual genius than a participatory and collaborative process, a taking part in the larger unfolding of evolution and ecology— and in architecture this involves collaboration with engineers of various sorts, as well as with experts in the study of the natural world and, perhaps also, human nature and cultures. The resulting buildings are designed not as isolated objects but as a nexus of interactive processes with their surroundings. Designs emerge from studying the interactions with every aspect of their setting, from cultural context and local building traditions to climate, geology, hydrology, ecology etc. Such a design process often draws on up to the minute forms of scientific survey as well as leading-edge engineering which, in turn, depends upon state of the art computer modeling and predictive analysis. These studies feed rather than constrain creativity, grounding it in a myriad of novel factors to be resolved and synthesized, while the participative ethos ennobles rather than diminishes it. Rather than being regressive, such an approach is very much in tune with major changes in the world view that science and the computer are affecting within our culture.

Nevertheless, green design has much to learn from the past, although there is nothing regressive about this either. Most major cultural epochs, such as that beginning with the Renaissance, are born in part from a rediscovery and reinterpretation of history. Just as the quest for sustainability is making us more receptive to the wisdom of other and earlier cultures, so green design can learn much from buildings of the past. Victorian buildings can be very sophisticated in their ventilation and heating systems, so it is little surprise that some recent green buildings in England (including some of those by Michael Hopkins) should rather resemble them. But, for the ways in which they are beautifully embedded in and adapted to their context and climates, it is usually vernacular buildings that are particularly instructive—both the vernacular of the local region and that of comparable biomes, those regions elsewhere on the earth with similar climate, vegetation and geology. Buildings of the past or other cultures are not to be copied slavishly, but rather studied so that their lessons and design devices might be reinterpreted to suit current construction technologies and lifestyles. The more recent past has much to teach us too, and not just in terms of negative examples. Many now forget that modern architecture grew in part from proto-green ideals, including the goal of creating community (most explicitly, and not always successfully, in the social housing built in Europe throughout the twentieth century) and of living in harmony with nature, whether this manifested in the physical form of the building or in the healthy open-air style of life it made possible.

Modern architecture always consisted of several streams, each of which had its own sources, which would flow independently, then join together to later branch apart again, thus creating a rich variety of intermingling approaches to design. If one proto-modern stream created an abstracted classicism, another advocated a return to Gothic architecture's direct expression of structure, construction and function. If one stream endorsed and celebrated industrial components and advanced technology, another lamented industry's despoliation of nature and its dehumanizing work conditions. John Ruskin, who has been called the father of ecology, was the key figure in the advocacy of the Gothic and the rejection of the negative aspects of industrialization that were already very visible in England. He saw the necessity of pursuing what we now call sustainability more than a century and a half ago. Thus in the "Lamp of Memory" (from *The Seven Lamps of Architecture*, 1849) he stated its moral imperative in terms that, except in language, are strikingly contemporary: "God has lent us this earth for our life; it is a great entail. It belongs as much to those who are to come after us, and whose names are already written in the book of creation, as to us; and we have no right by anything we do or neglect, to involve them in unnecessary penalties, or deprive them of benefits that it was in our power to bequeath."

Ruskin inspired the Neo-Gothic movement and the Arts and Crafts that followed it, and thus much of the ethos of what became modern architecture. (The young Charles Eduoard Jeanneret, who was to become Le Corbusier, was profoundly influenced by Ruskin, as was Frank Lloyd Wright.) Ruskin's own architectural ideals were realized most directly in Deane & Woodward's Natural History Museum in Oxford (1855), in the creation of which he played a very active role, both as an advisor to the architects and as the patron who paid for much of the carved ornament. The museum is in an eclectic Gothic style, its main space an exhibition hall flooded with natural light through a roof of glass tiles supported by a Gothic-inspired structure of cast and wrought iron. The ornament of the structure and the carvings around the windows meticulously record local flora and fauna, and the colonnettes of the galleries overlooking the hall are each of a different British stone. The whole building and its contents are both an education about and exaltation of nature. Once again much admired today, it was—along with Ruskin's writings—one of the original inspirations for two streams that have flowed through modern architecture.

The more vigorous and constant of these streams is what has come to be known as the biomechanical strand of modern architecture, whose origins include also the designs and writings of architects like Viollet-le-Duc. Rejecting the eclectic architecture of the preceding era, this initially sought more authentic expression in emulating the close fit between form, structure and function found in organisms, machines, or some formal fusion of these. After Neo-Gothic this stream flowed sinuously through Art Nouveau to eventually become the Vitalist strain of British High-Tech (with its exposed skeletal structures and knuckle-like joints, and shapely exoskeletal carapaces) and Renzo Piano's quest to bring an organicized technology into harmony with nature. Flowing more intermittently were the various tributaries of organic architecture, including that of Frank Lloyd Wright, which partly sprang from the Arts and Crafts and sometimes flowed close to the biomechanical and sometimes branched far away from it. Hence Wright's buildings were often built of local stone, nestled seamlessly into their sites and adapted to the local climate; yet they also made discriminating use of advanced construction and the technology of their times. Today, besides looking at vernacular architecture, with all its myriad passive devices of climate control, to learn about green design, modern architecture's biomechanic and organic streams (as well as others not discussed here) leave a rich legacy to reappraise and resurrect. It is hardly surprising that heirs to the biomechanical stream such as Foster and Partners,

Michael Hopkins and Partners and the Renzo Piano Building Workshop are among the leading exponents of green design.

The reasons for the withering of modern architecture's green ideals are too various and too complex to be more than, touched upon here. Most are not particular to modern architecture, but to modern culture generally. These include the whole restless ethos of capitalism with its constant drive to develop and redevelop, and the assumption underlying all industrial culture (capitalist, socialist or whatever) that growth is the ultimate good; the primacy of profit and the market, and the lack of value attached by economics to such 'natural capital' as clean air, fresh water and much else of nature's bounty upon which we are totally dependent; the tendency to short-termism in politics, planning and profit accounting; the tendency to reductive and instrumental thinking that tends to be so abstracted and narrowly focused as to miss the bigger picture; and the hubris brought by wealth, power and ever-improving technology that assumes that these can solve anything. In architecture, the critique of industrial society for its excessive exploitation of nature and humans, which had underpinned much early modernism, evaporated as modern architecture was co-opted, most willingly on its part, by commercial and governmental clients. These used the rhetoric of Functionalism to excuse the reduction of architecture to mere utilitarianism, minimal space standards and cheap construction. Buildings had value only in terms of efficiency of function (measured in profitable returns), and for as long as they functioned (remained profitable). Architecture shriveled into an autistic lack of concern with any larger issues. Compounding all these problems was the notion of an "international style"—that came with an equally destructive town planning accomplice, the Charter of Athens, promulgating mono-functional zoning—that virtually legitimized buildings quite unadapted to place and climate. Ironically, then, Functionalist buildings were often laughably inadequate in functional, climatic and constructional performance when compared with their nineteenth-century predecessors.[1]

With the introduction of modern industrial processes and products, many features of everyday building technology have become environmentally pernicious. These include the tendency to level plots and raze all vegetation and local features when constructing tract housing; the use of toxic, polluting and non-biodegradable materials as well as water-profligate plumbing fixtures, appliances and sewage systems; the voracious consumption of non-renewable resources; and, perhaps most especially, air-conditioning and the changes it has brought, especially the total dependency on electricity, the generation of which is a prime consumer of fossil fuels and source of carbon dioxide emissions. Air-conditioning brought about the second international style: the ubiquitous glass box, erected everywhere and at home nowhere. Its sealed and tinted glass walls sever contact with and awareness of sensual nature; and its deep plan removes people yet further from the outdoors to be totally dependent, at all times of day, not only on air-conditioning but also on energy-consuming artificial light. Worse still, the open or glass-partitioned interiors, devised for easy supervision, tend to squash many forms of social interaction. Architecture has been reduced to little more than energy-guzzling packaging; the façade, relieved of such essential functions as providing ventilation and shade, is a mere skin, its design merely a matter of current fashion; and the interiors are devoid of anything that might be called a room or processional sequence of spaces.[2] Little wonder that architecture collapsed into the arbitrary decorativeness of Post-Modernism.

For many people, the first intimations that energy-profligate buildings were not really viable in the long term came with the Arab-Israeli wars of the seventies and the subsequent shortage, and price hikes, of oil. Initially this provoked a considerable response in the United States, resulting mainly in experimental low-energy houses—many self-built by counter-culture sympathizers, and usually inelegantly rough and ready in form and construction—along with some larger complexes, such as those commissioned by counter-culture hero Governor Jerry Brown of California. However, oil became cheap and abundant again, and (apart from a few notable exceptions such as the Intelsat head-quarters in Washington by Australian architect John Andrews), architects working in—the United States and their clients ignored the green agenda. Instead green design, as noted earlier, was stigmatized as a marginal counterculture pursuit, associated with the most inurbane and unrefined of construction. In more recent years a trickle of green designs have been built (and there have been important experiments in related areas such as in green systems of sewage treatment, such as those using constructed wetlands). Although some of the resultant buildings are thoroughly convincing in green terms, many are aesthetically graceless, so doing little to overturn many architects' prejudices against green design. Worse, this reinforces a vicious cycle where by the better architects, who could create more formally accomplished buildings, remain wary of green design, so ensuring it retains its fringe status. Yet change is afoot and gathering momentum. With increasingly widespread and rapidly mounting interest in green design, and with mainstream architects and leaders in the profession beginning to show real interest, there is a gathering stream of promising projects on the drawing boards.

Nevertheless, the United States has fallen far behind other parts of the world, particularly Europe. There the development of green design has followed a very different course. The oil crisis of the seventies provoked some experiments in low-energy design, but not as many as in the United States. Since then, however, things have steadily escalated and more and more green buildings are being erected, mostly in northern Europe, though now in the Mediterranean countries as well. The architects now pioneering green buildings include some of Europe's foremost, such as Renzo Piano, Norman Foster and Michael Hopkins, as well as those beginning to win international acclaim such as Thomas Herzog and Françoise Jourda. Green buildings are far from a fringe pursuit, and are increasingly demanded by the most mainstream of clients. Indeed, for a building to have opening windows, fresh air, natural light and no air-conditioning conveys prestige. Such buildings include numerous flagship corporate headquarters, such as the skyscraper by Foster and Partners for Commerzbank in Frankfort (p. 46), and the far from stereotypical glass box by Webler & Geissler for the Gotz headquarters in Würzburg (p. 62), both in Germany. Even the most prestigious of governmental buildings are now green, such as Germany's reconstructed parliament, the Reichstag in Berlin (also by Foster and Partners), and Portcullis House, Michael Hopkins and Partner's building for members of Parliament, as well as one of the first of the current generation of green buildings, the European Investment Bank in Luxembourg by Denys Lasdun and Partners from the early 1980s.

This difference between the United States and Europe has been attributed to several things, all of which probably contribute to it. Fuel is more expensive in Europe, so the cost savings of energy efficiency are more pronounced. Utility companies and energy lobbies have not been able to suppress debate and information, and exercise as much power over policy, as they have in the United States. Some European countries have had strong clerical and office worker labor unions who demanded the physical work conditions their members prefer. A greater proportion of European businesses, both large corporations and smaller companies, build and occupy their own premises (rather than those of speculative developers) and so have direct interest in pleasant work conditions, diminished running costs and the benefits of a stable, happy and productive work force.[3] Much of Europe has a more temperate climate than large parts of the United States, and so it is easier to ensure comfortable conditions without resorting to air-conditioning. Europe consists of relatively small countries that demand action when pollution from a neighboring country defiles their rivers or kills their lakes and forests—in contrast to what can be resigned acceptance of home grown pollution. Leading European architects have generally been more concerned with social and technical issues, function and performance, than many of the most prominent American architects, whose concerns are more with form and theory. Engineers play a larger and more collaborative role in the design of buildings than in America, their proportionally larger fee affording them twice as much creative design time. In parts of Europe the banking system and funding of construction is somewhat less geared to the short term than in the United States, and so is less inhibiting of the long-term thinking and accounting essential to dealing with environmental issues. Generally too, European building, planning and tax codes are less likely to inhibit green buildings than some of those in the United States.

Also, in Europe in the last decade or so, green design has been given further impetus by the active encouragement of national and local governments (especially in Germany), and that of the European Union, which is firmly committed to working towards sustainability in general and energy efficiency in particular. Already in the seventies and eighties much of Europe introduced stringent energy-saving building codes, starting with such things as insulation standards. Then more recently the European Union, through its Joule 1 and Joule 2 programs, has paid for the research and experiment costs involved in developing and testing the designs of some high profile green buildings, reasoning that architecture generally would benefit from the widely publicized results. This is complemented by, among other initiatives, the Thermie program whereby the European Union contributes to the costs of experimental projects, in particular paying for the installation and testing of untried and still uneconomic environmental technologies. The rationale is twofold: that testing any technology promising considerable environmental benefit is an urgent imperative; and that some proven technologies would quickly become economic if demand was raised sufficiently to bring economies of scale, and further refinements, to their manufacture. The photovoltaic cells on both the Jubilee Campus of Nottingham University by Michael Hopkins and Partners (p. 74) and the Mont-Cenis Academy in Herne-Sodingen by Françoise Jourda Architects (p. 90) were paid for by Thermie grants. In the latter case, this helped make viable a nearby photovoltaic factory, with initial orders big enough to ensure its product will soon be sufficiently cheap to be economically viable (in terms of how quickly the cost of installing the photovoltaic is offset by the savings in utility bills).

Although Germany, Britain, the Netherlands and the Scandinavian countries are all independently committed in principle to achieving sustainability, and are each making considerable progress in reducing green house gas emissions, Europe still has a long way to go. By contrast,

though, the United States resolutely refuses to address the issues. In 1997, under the Kyoto Protocol, the industrialized countries committed themselves to reducing greenhouse gas omissions by at least 5% below 1990 levels by 2008-2012. But, as the European countries recognize, this is far too little, far too late. (Scientists calculate that to avoid the worst consequences of global warming, cuts of 60-80% are required immediately.) Hence Germany's 1999 carbon dioxide emissions were already 15.3% below those of 1990 and it will easily exceed its own target of a 25% reduction by 2005. For all six gases covered by the Kyoto Protocol, Germany has already achieved an 18.5% reduction towards a goal of 21% by 2008-2012. Britain has cut its greenhouse gas omissions by 13%. It has now introduced new measures estimated to achieve a 21.5% reduction in carbon dioxide emissions by 2010. All this is being achieved relatively painlessly, mostly through the introduction of more efficient technology and careful design. By contrast, the carbon dioxide emissions of the United States (with less than 5% of the world's population and producing 25% of all carbon dioxide emissions) have increased by 21.8% since 1990. Instead of reducing its emissions in accord with the Kyoto Protocol, the United States proposes purchasing, and having considered as its own, portions of the reductions achieved by other countries. The United States also scuttled the ratification of the Kyoto Protocol in The Hague in November 2000 by insisting that the increased capacity to absorb carbon dioxide achieved through tree planting be considered as equivalent to a cut in emissions—despite the warning by many scientists that tree planting in temperate or cold climates is likely to increase global warming in the medium to long term. Even more recently, the new Bush Administration has announced that it has no interest in pursuing the Kyoto goals. The United States is thus abnegating totally its role as by far the most influential and emulated country in the world, and committing millions of people to a miserable future. It is foregoing the great benefits in improved quality of life that could have been achieved with little effort beyond the exercise of the imagination.

The global environmental problems caused by the United States, however, result from far more than the irresponsible example it sets and its emissions of greenhouse gases and other pollutants. It also consumes far too much of the earth's resources generally, not just too much fossil fuels. If every person now living were to consume as much as the average United States citizen, it would require, by the most conservative estimates, at least another two of the planet Earth to support them. Americans, like everybody else, need to urgently reappraise their priorities and reconsider what they really want, now and most importantly, for the future. This should provoke a profound shift—away from the simple-minded consumerist pursuit of more, toward the far more discerning quest to obtain more of only what we really want—of what would make us truly and deeply happy.

It should be clear by now that green design, though not dauntingly difficult, cannot be achieved by any simplistic or formulaic approach: no single approach is likely to be adequate, let alone appropriate or even applicable, to all situations. Green design goes far beyond merely specifying efficient 'green' products, such as insulation, low-emissivity glass, water-conserving toilets, super-efficient mechanical equipment and non-polluting materials; and beyond also using replenishable, recycled and recyclable materials, recycling all rain and 'grey' water and planting on roofs. Green design both influences the basic design parti of a building, especially the cross-section and the elaboration of the outer envelope, and transcends mere energy efficiency and the minimization of pollution. Instead it must attend to a whole range of

matters from the technical and ecological, to the economic and social, including even the cultural and spiritual.

This book spells out clearly both how broad is the span of these issues, and what they are exactly. Ten "shades," or aspects, of green are discussed in detail in the following section of this book: 1— ow energy/ high performance; 2— replenishable sources; 3—recycling; 4—embodied energy; 5—long life, loose fit; 6—total life cycle costing; 7—embedded in place; 8—access and urban context; 9— health and happiness; 10—community and connection. This list is intentionally somewhat hierarchic, starting as it does with narrowly technical issues, many of them quantifiable, and leading up through contextual and urban issues to largely qualitative socio-cultural ones. As listed, then, the issues could almost be seen as forming a conceptual ladder of steps of ever-broadening concern, with each step having to be considered in turn if pursuing a fully green approach to design. However, the position on the ladder does not indicate the relative importance of the concern. For example, as buildings become more energy efficient, embodied energy accounts for a proportionally larger part of the total energy invested in a building's total life cycle; yet embodied energy will probably remain less critical to the larger quests for green design and sustainability than the other issues listed directly before and after it in the sequence.

Two key issues are missing from the first set of ten shades: the roles of collaboration and of the computer. They are omitted both because they do not apply to all kinds of green design, and because rather than fitting into the hierarchy they relate to all steps of the ladder. With relatively traditional building types, such as houses and housing, satisfactory green design can be achieved by little more than updating vernacular building types and drawing on the accumulated empirical experience of the architect. But with larger and somewhat newer building types, for which there may be no vernacular precedent, devising a fully green design often requires a greater range of skills than any architect can provide. Engineers of various sorts make crucial contributions to the design, as might ecologists, hydrologists and geologists, horticulturists and landscape architects and various other specialists—including those with intimate knowledge of local climate, materials and crafts. To best synthesize their inputs, all these disciplines should contribute right from the beginning, some specialists (particularly the structural and mechanical engineers) participating by playing a fully creative role in a highly collaborative design process. This is relatively common practice in Europe. But learning to work in this way, with engineers as partners in the creative process, may be a challenge for U.S. architects.

The computer plays many roles in green design: during the design process it aids synthesis and prediction, and in the final building it constantly monitors and adjusts conditions. In early stages of design, the inputs of the various specialists might first take the form of surveys, whose findings might be weighted relative to each other and synthesized with the aid of the computer. Later, predictive modeling by computer and other forms of testing are also crucial. With conventional buildings, uncomfortable ambient conditions can be rectified by cranking up the air-conditioning. But because green buildings probably have no air-conditioning or other energy-guzzling equipment to fall back on, they have to be engineered with a precision made possible only by the computer. Ensuring that the building under design will remain comfortable, even after prolonged bouts of extreme temperatures, necessitates all kinds of predictive modeling and testing, studying fluctuations in temperatures, light conditions, humidity and so on. The completed buildings also depend heavily on computers. With the aid

of myriads of sensors, the computerized 'building management system' monitors conditions inside and out; and with the aid of sophisticated software, that might use such things as 'fuzzy logic' and 'neural networks', it responds by constantly adjusting (with the aid of low-energy electric motors) the various mechanical devices that control the sun-shading, ventilation, flow of cool ground water in ceiling panels, and so on. In short, the computer makes it possible for buildings to be designed and to function as integral to—as a set of processes in reciprocal interaction with—its environment.

Increasingly, buildings under design are studied for their impacts on an area as much as a mile around them. The impacts on surrounding streets and buildings of, for instance, new wind patterns and pressures, or of exhausts such as steam and the condensation and increased temperatures these may cause, are modeled and analyzed. In part this is in acknowledgement that, because it is now possible to do this, others would be entitled to sue architects and clients for any disturbances caused by new buildings.

A third issue omitted from the first set of ten shades is the redefinition of comfort standards. This was omitted because it is probably only going to be an issue in the short term, before being widely accepted (as it already is in Europe). Consistent with industrial culture's obsession with standards and standardization, comfort used to be defined in terms of constant levels of temperature, humidity, air changes, light and so on. Yet research, no longer recent but previously little noted, about people's actual experience of comfort proved the common-sense notion that in winter, when more warmly dressed, people feel comfortable at lower temperatures than in summer, when more lightly dressed.[4] This might seem not only an obvious but also a rather trivial observation; yet complying with its implications alone can result in enormous energy savings, simply by lowering the levels of energy-hungry winter heating and summer cooling. Other studies have shown that people prefer the sensual awareness of their bodies brought by gently fluctuating conditions (that open windows offer, for instance) to the homeostatic and anesthetizing constant conditions once considered to be ideal.

The buildings presented in this book and the exhibition it documents were selected to prove this diversity of potential solutions to the same green issues—and so also to show that resolving these is no straitjacket but rather a stimulus to creativity. This heterogeneity shows that there is no such thing as a green aesthetic, while the high architectural quality of the works shows that green buildings need not be synonymous with ugly and unrefined design. Some schemes, such as the Cotton Tree Pilot Housing by Clare Design (p. 56) and the four North American houses (by Lake/Flato, Rick Joy, Fernau & Hartman and Brian MacKay-Lyons (p. 102–111), are examples of the local vernacular reinterpreted for today; others, such as the Commerzbank tower and the Jubilee Campus, are the products of leading edge engineering and predictive computer modeling, by mechanical engineers Roger Preston and Partners and Ove Arup and Partners, respectively. Crucially, all the buildings illustrate well the enhanced quality of life promised by green design.

To facilitate instructive comparison, contrasting buildings of similar functional type were chosen. Thus there are two office buildings, both of them corporate headquarters in Germany that were purpose-built for owner-occupier clients. The tall tower of the Commerzbank headquarters is one of a cluster of towers in downtown Frankfurt; the Götz headquarters is a low-rise glass box—that at first glance seems only a refined version of what might be expected in its suburban, business park location. Yet despite these contrasts, the buildings have much in common. Both are naturally lit and ventilated through double-layered outer walls and internal gardens that bring the sensual presence of plants into the buildings while also serving as social foci. In the Götz headquarters this garden is in the central atrium whose glazed roof lifts and slides aside in summer; the Commerzbank tower's equivalent are the sky gardens that spiral around the central shaft that some people refer to as an atrium. Both buildings are cooled by chilled ceilings, and share a common spirit reflected in a very similar approach to details, as seen in particular in the glass balustrade. Both—with their bright and airy, naturally lit and ventilated, verdant interiors with ample provision for socializing—are green for more reasons than energy-efficiency; those very elements that are intrinsic to their energy efficiency (the sky gardens and central atrium) also enrich the buildings' social, sensual and spiritual dimensions.

The two housing schemes form another contrasting pair. Beyond concerns with energy efficiency and community, however, these have nothing in common. Because of their very different locations, this is as it should be: Slateford Green by Andrew Lee with Hackland + Dore is in Edinburgh, Scotland (p. 96), a city of cold northern European winters; the Cotton Tree Pilot housing, by the husband and wife team of Lindsay and Kerry Clare of Clare Design, is on the semi-suburban Sunshine Coast of tropical Queensland, with its year-round heat and humidity. Slateford Green is a single building, a looped wall of contiguous units wrapped around a single central court that, together with the conservatories overlooking it, is designed as a winter sun trap protected from the cold winds off an estuary of the Firth of Forth. Some of the Cotton Tree dwelling units also form a wall along the scheme's southern edge. But large gaps in the wall let the cooling breeze blow through it and between the units clustered on the rest of the site, as well as through the units themselves, while overhanging roofs shade windows and balconies. Here the cars are not kept to the perimeter, as at Slateford Green; instead motor courts penetrate the site. The motor courts, and the access routes from them to the dwelling units, form a dispersed communal realm quite unlike that of the singular central garden (ringed by private gardens) in Edinburgh. Yet thanks to careful design, both housing schemes promise a lively community life in an outdoor realm largely landscaped with indigenous plants (that tend to be better adapted and so 'greener' than imported ones).

Three institutes of higher learning are probably the most startlingly radical of the selected designs. Each looks very different from the other two, or indeed any other building. Each also exploits very different means to modulate its internal conditions and achieve energy savings. Almost disconcertingly strange, yet thrillingly so too, is the Minnaert Building by Neutelings Riedijk on the Uithof Campus of the University of Utrecht in the Netherlands (p. 84). Inside its rust-colored, sprayed concrete outer coat, wrinkled to express that it is only a skin, is a cavernous central hall. Half of this is occupied by a pool into which rainwater plunges, to then be circulated around the building to conduct away the heat given off by computers and occupants. The Akademie Mont-Cenis, a training center for local government employees in Herne-Sodingen in Germany, was built as part of the regeneration of the ex-rust belt of the Ruhr region. Standing in a new park over an abandoned coal mine, it resembles some cross between an enormous greenhouse and a contemporary temple, combining two design idioms that are usually polar opposites: crisply minimalist high-tech glazing and funkily rustic tree-trunk columns. Designed initially for a competition by Jourda & Perraudin in association with Ove Arup and Partners, and executed by Françoise Jourda in association with Hegger Hegger Schleiff, it offers inside a Mediterranean microclimate where students can enjoy a hedonistically semi-outdoor life in the elongated piazza between the residential and academic blocks. Another competition-winning design, the Jubilee Campus, by Michael Hopkins and Partners with engineers Ove Arup and Partners, is also built on what had been a derelict industrial site on which the architecture now sits in verdant landscaping. Here building, landscaping, ventilation system and the wind itself all work together as an inextricably interwoven whole to achieve a very complete and efficient green scheme that includes many devices that could be adapted and applied widely elsewhere. With its lakeside promenade and glass-roofed atria, this is a complex every bit as tailored around pleasant socializing as the Mont-Cenis academy.

Four North American houses together comprise one of the second set of ten shades. (A solitary house could hardly have represented the whole continent.) Each is in a different climatic region, and each updates or reinterprets the local vernacular to create houses that look at home in their settings and are well adapted to the climatic conditions. The Palmer House by Rick Joy, outside Tucson, uses the thermal mass of thick adobe walls to absorb heat and stabilize the large diurnal temperature fluctuations of the Arizona desert. The South Texas Ranch House by Lake/Flato is in a somewhat less arid part of Texas and combines the patio format of local tradition with a freestanding pavilion that opens up to breezes off the adjacent river. The Westcott-Lahar house by Fernau Hartman is fragmented into elements that zigzag across an upper corner of the large site onto which the various rooms open. The back wall of the house is built of straw bales whose snug insulation is emphasized visually and psychologically by the resulting deep reveals to the small windows. These windows contrast with the large glass doors opposite that open onto the garden and terraces. The strong, simple form of the Howard

House by Brian MacKay-Lyons commands its seaside site in West Pennant, Nova Scotia, and adopts the matter of fact approach to available materials found in the local vernacular. All four houses remind us that it is only in relatively recent times that architects, and even then far from all of them, have chosen to ignore so disastrously climate and context.

Only two of the buildings presented are solitary examples of their functional types. These are Hall 26 by Thomas Herzog, an exhibition hall in a trade fair complex outside Hanover, Germany (p. 68), and the Beyeler Foundation Museum by the Renzo Piano Building Workshop, outside Basel, Switzerland. Yet they share some energy-saving strategies: both admit abundant natural light through their roofs; and both exploit the way air stratifies and rises slowly with increasing temperature, thus obtaining stable conditions in the lower portions of their interior volumes despite fluctuating conditions closer to the ceilings. Both buildings were selected to illustrate other points as well. The whole form of the exhibition hall, especially its longitudinal section, is an exact correlate of the efficient way it admits and reflects light as well as channels and exhausts the rising air. The museum exemplifies beautifully the ideal of a building embedded in its place so well as to seem to have grown from rather than been imposed upon its setting. Both buildings, which share an aesthetically satisfying crisp precision of detail, were also chosen because they are by architects who have designed other significant green buildings.

As well as for the reasons outlined above, the buildings were selected to show work from a range of architects: from established international figures to relative unknowns; and from those who have established a reputation for green design to those who are still relative novices in the field. All four of the established international figures—Norman Foster, Thomas Herzog, Michael Hopkins, and Renzo Piano are

leaders also in green design. Just as instructive as comparing green buildings of similar functional type, is to study how each of these architects has explored and evolved different approaches to green design in successive buildings (an exercise recommended to all readers). Significantly, a number of the key green buildings by each of these architects, except for Hopkins, are in Germany, and the experience of building in that country had either led initially to or deepened each architect's embrace of the green agenda.

Thomas Herzog first embraced green design in the seventies, with some fine private houses. Since then his green buildings have included, as well as more houses, the Linz Design Center, an exhibition hall in Austria. For its roof he developed a special sandwich glazing that admits a high proportion of sunlight while excluding direct shafts of sun. Close to Hall 26, he has also built an energy-efficient office tower and a set of gigantic timber parasols under which shelter smaller exhibition pavilions. The office building has a double skin of glazing, a now fairly common solution on German green buildings—and taken to an extreme of sophistication on the Götz Headquarters by Webler + Geissler. The parasols push timber construction to new limits and are supported by what had been three hundred year-old conifers. Felling huge old trees may not strike some as green, but Germany's foresters insist that it is, because such mature trees no longer absorb as much carbon dioxide as the growing trees that replace them.

Foster and Partners' first truly green buildings are in Duisburg on the Ruhr and, among other devices, again exploit double layers of glazed façade. After the Commerzbank came the reconstruction of the historic Reichstag in Berlin that once again houses the parliament of reunified Germany. Conspicuous outside is the new steel and glass dome that rises from the roof, as light and optimistic as the original dome seemed heavy and oppressive. Hanging down from the dome into the assembly chamber is the 'light reflector'. Its mirrored facets reflect natural light down into chamber while warm stale air is drawn up and out through its funnel-like form. As the dominant feature of the chamber, the light reflector hovering above the heart of Germany's government serves as a resonant symbol of its commitment to green ideals. As further green measures, the building powers its own electric generator with vegetable oils (releasing less carbon dioxide than captured by the plants providing the oil) and stores for recycling both warm and cold water in chambers insulated deep below the earth. Destined to be yet another of Foster's seminal green buildings is the Swiss Re office tower to be built in the City of London. This is an evolution of the Commerzbank designed for a corporation that has already commissioned and occupied green buildings in its native Switzerland. From this they know the advantages of green design lie in improved quality of life and social interaction, and so improved internal communications, as much as in diminished running costs and environmental stress. Instead of a hollow shaft, Swiss Re will have a conventional core. But corkscrewing around its perimeter will be a series of naturally ventilated shafts with gardens on their stepping ledges. Meetings on these will be visible and inviting to those on the floors above and below them, so aiding the ease of communication that the clients value highly for the efficiency it brings. Like the Commerzbank, this is a design that has evolved in part by modeling its impacts upon an extensive area around it—in this instance by the mechanical engineers BDSP.[5] The rounded form of the building minimizes air turbulence and ground-level gusting, while the differing wind pressures around the circumference move air through the corkscrewing shafts without need for mechanical backup. In this case, the chillers were repositioned during design when computer modeling proved that

they would sometimes have caused condensation on buildings some distance downwind.

Although it uses similar devices to those already explored at the Cité Internationale in Lyon, France, the Renzo Piano Building Workshop's first rigorously researched (with funding from the Joule 2 program) and designed green buildings are the office blocks of the large Potsdamer Platz scheme in Berlin. Shallow office floors, and an elongated glass-roofed atrium in the largest block, allow these buildings to be naturally lit and ventilated through opening windows. To break the force of the wind against these windows, and to intercept rain so that the windows may be safely left open at night to allow night time purging of summer heat, the taller facades are protected by an outer layer of adjustable horizontal glass louvers. In summer these can be opened to admit breezes, or be semi-closed to channel a stack-effect upward draft that draws cool air up the faeades and into the open windows. In winter the louvers are closed to trap a layer of air, intermediate in temperature between inside and out, against the building. This forms both an insulating layer and a source of warmed air for ventilation. All the offices include perimeter heating, and can have ceiling panels chilled by groundwater. These are used only in the extremes of winter and summer, respectively, when mechanical ventilation is available too. All rainwater is captured and used on planting, and grey water from hand-wash basins is used to flush toilets. The whole scheme is green in other ways too. It is a dense, mixed-use urban scheme (where people could live near to work, shops, etc.) that is also highly accessible by public transport—although these aspects are offset somewhat by the generous car parking provision and the large triple-level shopping mall dedicated to conspicuous consumption. The buildings are faced in terracotta that mellows with age and weathering and are flexible in plan, so exemplifying the ideal of long life, loose fit. Also, during construction most materials were brought to site by canal to minimize traffic congestion and pollution.

Another of Piano's seminal green buildings is the J. M. Tjibaou Cultural Center outside Nouméa, capital of the Pacific island nation of New Caledonia. It was built to conserve the artifacts and revivify the culture of the local Kanak people. Here enormous basket-like carapaces of curving, vertical timber ribs and horizontal wood slats cluster along a ridge between the open ocean and a sheltered lagoon and protect the center's main spaces. The prevailing winds off the sea sigh in the slats of these cases (as the architect calls them), which recall the spirit rather than exact forms of traditional Kanak huts (case in French). With the small scale and the optical vibration given by the slats and their varied spacing, these cases also establish a remarkable visual affinity with the surrounding vegetation that is venerated by the Kanaks. Thus the building is thoroughly embedded in place, in both its physical setting and in the culture it evokes in what is also a very contemporary design language. By adjusting banks of glass and wooden louvers behind the slats, these spaces are naturally ventilated no matter the strength and direction of wind—whether still days, when stack-effect ventilation is induced, or during typhoons from either ocean or lagoon. But for all their highly allusive poetry, these cases are also extravagant gestures; they stand as tall as the nave of a Gothic cathedral yet enclose smallish, single-story spaces. A Gothic cathedral, however, is also a single-story space; perhaps the best way to think of this building is as a temple to a new sensibility that has rediscovered a reverence for nature and for the traditional cultures that worshiped and lived in easy harmony with nature.

The Jubilee Campus was not the first milestone of green design by Michael Hopkins and Partners (another architect whose designs have benefited from research sponsored by the Joule 2 program). Another, also in Nottingham and also a commission won in competition, is the earlier Inland Revenue complex. To achieve long life loose fit, this is solidly built in brick and concrete and broken up into smaller blocks that can be leased independently if the original client no longer needs all of them. These blocks consist of a constant narrow section, allowing natural light and ventilation, folded around courts. Brick piers and light shelves both shade the windows from direct sunlight and reflect light deep into the offices; the brick piers and precast concrete ceilings also provide the thermal inertia to stabilize temperature fluctuations. Projecting from the corners of the blocks are cylindrical stair towers clad in glass blocks. As the sun warms these, the air inside rises by stack effect, its rate of flow—and so that of the air drawn through the offices to replace it—being adjusted automatically by the computer controlled raising and lowering of the stair tower's roof, below which the air is exhausted.

Portcullis House, opposite London's Houses of Parliament and providing offices for some of its members, was designed before but completed after the Jubilee Campus. Built of stone and precast concrete, this too has a high thermal inertia and is built around a court so that it can all be naturally lit. But because bounded by heavily trafficked, noisy and polluted streets, it does not have opening windows. Instead fresh air is drawn in above the roof ridge to be distributed down the facade and admitted through grilles in the raised floor. Stale air from under the ceilings is drawn up the facade to exhaust through heat exchangers in cylindrical chimneys that give the building a silhouette very similar to those of its neighbors.

Buildings such as those presented in this book and others by the same architects, as well as those by other architects exploring similar themes, are stepping stones to the future. They resonate with the epochal changes sweeping through our culture as evidenced in the emergence of the new sciences, the computer and how it is transforming our world and our understandings of it, and our deepest values. Yet this seems not to be fully recognized in the U.S., particularly in architectural schools and in the journals. This is because these tend to cling to a waning cultural paradigm or are seduced by what is merely a transitional one. The dying cultural paradigm is that of four hundred years of modernity, which emerged along with science and was consolidated during the Enlightenment. The single big idea underlying this paradigm, and defining its sense of reality, was that there is an objective reality, external to and unaffected by any observer. As this paradigm's weaknesses have become more evident and its fundamental flaws exposed (by, for instance, quantum mechanics) we have entered the age of post-modernity in which the cultural pendulum has swung to the opposite extreme and the emphasis is on subjective interpretation. In this cultural paradigm, in which everything is relative rather than objective, reality is a subjective projection or consensual hallucination. However, this paradigm is not the long-term successor to modernity; instead it is both its suppressed and contrary flip side and a hyper- modern caricature of it. As caricature, it severely exaggerates some of modernity's greatest weaknesses, such as its propensity to excessive abstraction, and so alienation, arising in part from a lack of grounding in sensual experience. Underpinning the long term cultural paradigm emerging to replace modernity is a notion of reality that is both larger than and external to us, and yet is shaped in some degree by human participation. Reality in this paradigm is multi-layered and constituted by the unfolding and interlinked processes of cosmological, natural, socio-historic and personal evolution—a hierarchy whose lower levels we help create

through our perceptions and actions, and whose higher levels we might eventually impact upon benevolently as we understand them and apply these understandings.

Most contemporary architecture conforms to the modern or post-modern paradigm. As modernity loses its credibility, many architects are fighting a rearguard action by reducing what was once a more encompassing and ambitious approach to architecture down to narrow certainties, such as with form and material (minimalism), technology (high-tech) or a degraded view of function (commercial viability). Others are post-modernists, not necessarily because they quote from history (classical or modernist) to suggest symbolic or semiotic resonances, but because their works are largely devised as interpretations and illustrations of some spurious theory rather than being grounded in a larger web of realities. The architecture of the emergent new long-term paradigm must be born from an evolutionary and ecological perspective, to be good for both planet and people and grounded in the complex and sensual realities of place and lived experience. Ultimately this is the most compelling reason that green buildings are the inevitable, inescapable future of architecture.

*London, 2000*

## NOTES

1. Yet in many parts of the world, even the International Style was quickly adapted to suit to local conditions and climate and became an apt modern vernacular. Hence in tropical Brazil buildings arose on pilotis to better catch the breezes that wafted through their open, sun-shaded facades and the flowing spaces of their narrow, free-planned interiors.

2. This is not to say that air-conditioning will not remain a boon, and sometimes a necessity, especially in hot climates, polluted cities, art galleries and operating theaters. But careful design could ensure immense improvements in its energy-efficiency and eliminate the associated use of daytime artificial light, as well as ensuring that the air-conditioning be required only during extremes of heat and pollution. In fact, the tendency in some areas is to use mechanical ventilation only, rather than full air-conditioning, with chilled ceiling panels providing summer cooling.

3. Whether this claim is correct or not, it is certainly true that the first significant green buildings were almost all for owner-occupiers--although now even some speculative buildings are designed to be green.

4. Peter Buchanan, "Steps Up the Ladder to a Sustainable Architecture," *A+U*, no. 320, (May 1997), p. 6–13.

5. BDSP was formed by engineers who once worked for Roger Preston and Partners, who were mechanical engineers for the Commerzbank.

# BUILDING SOLUTIONS TO CLIMATE CHANGE

By the Pew Center for Global Climate Change

## INTRODUCTION

Energy used in residential, commercial, and industrial buildings produces approximately 43 percent of U.S. carbon dioxide ($CO_2$) emissions.[1] Carbon dioxide is the major greenhouse gas that contributes to global warming.

Given the magnitude of this contribution, it is essential that efforts to control global warming include an explicit focus on the buildings sector. This brief provides an overview of technologies and policies, examines current public and private initiatives to promote greenhouse gas (GHG) reductions in buildings, and makes recommendations for moving toward a climate-friendly built environment.

The United States has made remarkable progress in reducing the energy and carbon intensity[2] of its building stock[3] and operations in the last few decades. Energy use in buildings since 1972 has increased at less than half the rate of the nation's economic growth, despite the increase in average home size and growth in building energy services such as air conditioning and consumer and office electronic equipment. Although progress has been made, abundant untapped opportunities still exist for further reductions in energy use and emissions. Many of these—especially energy-efficient building designs and equipment—would require only modest levels of investment and would provide quick pay-back to consumers through reduced energy bills. By exploiting these opportunities, the United States could have a more competitive economy, cleaner air, and lower GHG emissions.

## THE CHALLENGE

GHG emissions from the building sector in the United States have been increasing at almost

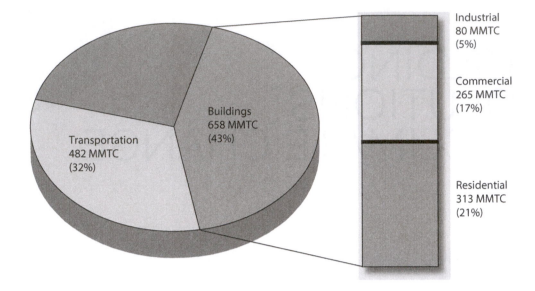

FIGURE 1. CARBON DIOXIDE EMISSIONS FROM FOSSIL FUEL COMBUSTION.

2 percent per year since 1990, and $CO_2$ emissions from residential and commercial buildings are expected to continue to increase at a rate of 1.4 percent annually through 2025. These emissions come principally from the generation and transmission of electricity used in buildings, which account for 76 percent[4] of the sector's total emissions. Due to the increase in household appliances and equipment that run on electricity, emissions from electricity are expected to grow more rapidly than emissions from fuels used on-site in buildings. In addition to the growth in demand for energy services within individual buildings, the U.S. building stock is also expected to double in the next 30 years.

Despite efficiency gains, emissions from the buildings sector are rising because the U.S. building stock continues to grow annually and the size of homes has increased significantly, which in turn increases energy requirements. Additionally, the range of electric equipment provided in buildings has increased significantly, especially air conditioning in the South and electronic equipment, televisions, and other "plug loads"[5] in buildings nationwide. Central air conditioning is now a standard feature of commercial and institutional buildings as well as 85 percent of homes in the United States, up from 34 percent in 1970. In order to compensate for this increase, more effort must be focused on increasing the efficiency of the buildings as well as providing affordable, low-carbon on-site electricity, and using waste thermal energy. Based on energy usage, opportunities to reduce GHG emissions appear to be most significant for space heating, air conditioning, lighting, and water heating.

The fragmented nature of the building sector poses additional challenges to promoting climate-friendly actions—distinct from those in transportation, manufacturing, and power generation. The design of effective policy interventions must take into account the multiple stakeholders and decision-makers in the building industry and their interactions. Major obstacles to energy efficiency exist, including:

- Insufficient and imperfect information (e.g., electricity bills bundle the consump-

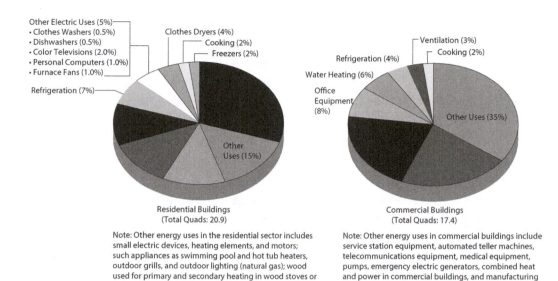

FIGURE 2. PRIMARY ENERGY CONSUMPTION IN RESIDENTIAL AND COMMERCIAL BUILDINGS, 2002.

*Source: Energy Information Administration, 2004. Annual Energy Outlook 2004. DOE/EIA-0383, p. 139–142, tables A4 and A5. EIA. Washington, D.C.*

tion of numerous end uses; at industrial facilities electricity use is charged to an overhead account.)

- Distortions in capital markets (e.g., electric utility profits are tied to sales in most markets, creating a disincentive for utilities to implement demand-side management (DSM) programs.)

- Split incentives that result when intermediaries are involved in the purchase of low-GHG technologies (e.g., incentives vary for the architect, builder, owner, and operator in terms of realizing cost savings from energy-efficient features of buildings.)

Most commercial buildings are occupied by a succession of temporary owners or renters, each unwilling to make long-term improvements that would benefit future occupants. Regulations, fee structures in building design and engineering, electricity pricing practices, and the often limited availability of climate-friendly

technologies and products all affect the ability to bring GHG-reducing technologies into general use. Some of these obstacles are market imperfections that justify policy intervention. Others are characteristics of well-functioning markets that simply work against the selection of low-GHG choices in the absence of mandatory GHG policy.

Despite these challenges, numerous individuals, corporations, communities, cities, and states are driving the implementation of "green" and carbon-reducing building practices in new residential and commercial development. Affordability, aesthetics, and usefulness have traditionally been major drivers of building construction, occupancy, and renovation. In addition to climatic conditions, the drivers for energy efficiency and low-GHG energy resources depend heavily on local and regional energy supply costs and constraints. Other drivers for low-GHG buildings are clean air, occupant health and productivity, the costs of urban sprawl, and stress on the electric grid that can be relieved by managing electricity demand.

## CURRENT DEVELOPMENTS IN GREEN BUILDINGS

In the absence of federal legislation on green buildings, or a comprehensive federal strategy to reduce GHG emissions, numerous stakeholders have begun taking actions that address the built environment's role in climate change. A combination of voluntary and mandatory measures taken at the local, state, federal, international, and corporate levels have all provided steps in the right direction. Whole Buildings Standards, for example, have provided a metric by which to compare buildings based on a variety of characteristics. Some of the most impressive progress is the result of communities and developers wanting to distinguish themselves as leaders in the efficient use of resources and in waste reduction in response to local issues of land-use planning, energy supply, air quality, landfill constraints, and water resources. Building owners and operators who have a stake in considering the full life-cycle cost and resource aspects of their new projects are now providing green building leadership in the commercial sector. However, real market transformation will also require buy-in from the supply side of the industry (e.g., developers, builders, and architects).

The term "green building" is used by a number of public and private programs to promote environmentally friendly construction practices. Most of these programs use labeling based on a point system to communicate to the market the relative environmental Furthermore, a number of progressive groups are spear-value of these practices. The standards currently serve a useful role in guiding stakeholders towards more climate-friendly green building practices.

However, further research is needed to better understand the life-cycle of GHG emissions of various building materials and appropriately account for them in building standards. Also, most standards are flexible enough to enable buildings to receive a "green" rating if they perform well on, for example, indoor air quality, even if they perform less well on energy efficiency and GHG emissions. This needs to change to ensure that buildings reduce their contribution to climate change. The U.S. Green Buildings Council (USGBC) has its Leadership in Energy and Environmental Design (LEED) standard,[6] which is perhaps the most well known whole building standard.[7] In addition to its well-established standards for new buildings, USGBC recently developed standards for building retrofits and for neighborhood development. There are several other whole building standards in addition to LEED: Model Green Home Building Guidelines, the Minnesota Sustainable Design Guide, and the Green Building Initiative (the Green Globes certification is administered through the Green Building Initiative). Currently 14 states[8] have adopted LEED as the standard for government buildings, and two states have adopted Green Globes.

Many states are going above and beyond traditional building codes by instituting their own green building standards. California and New York have both announced Green Building Initiatives. Governor Schwarzenegger has committed California to leading by example in improving the energy performance of existing and new State buildings by mandating that they reduce electricity consumption 20 percent by 2015. New York offers tax credits for energy efficiency measures and provides low-interest loans for building materials that meet LEED or other accepted green building standards.

Furthermore, a number of progressive groups are spear-heading several initiatives to

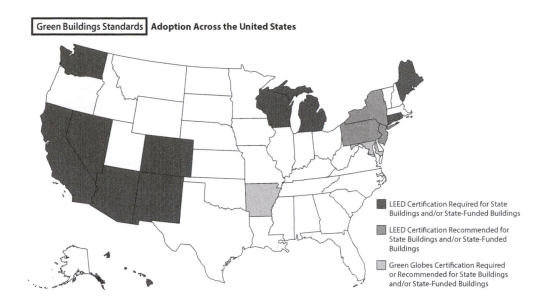

FIGURE 3. GREEN BUILDINGS STANDARDS ADOPTION ACROSS THE UNITED STATES.

address this issue. In January 2006, the group Architecture 2030 publicly issued the "2030 Challenge" (www.architecture2030.org). Since then, the American Institute of Architects has adopted the "Challenge" calling for architects and others in the buildings industry to reduce GHG emissions in new and renovated buildings 50 percent by 2010 and to make all new buildings "carbonneutral"[9] by 2030. In June 2006, the U.S. Conference of Mayors unanimously passed Resolution #50 urging cities across the country to adopt the "2030 Challenge" for all buildings, and setting benchmarks and timelines to achieve the goals. In July 2006, the International Council for Local Environmental Initiatives (ICLEI) North America unanimously supported the "2030 Challenge" and embedded its targets in ICLEI's "Statement of Action." New Mexico Governor Bill Richardson issued an executive order in 2006 committing the state to the 50 percent reduction target for new state buildings. And ASHRAE, the USGBC, and the Illuminating Engineering Society (IES), with input from the

AIA, are in the process of developing ASHRAE #189, a new high performance building standard that will incorporate similar targets.

On the international scene, the World Business Council for Sustainable Development (WBCSD) launched an Energy Efficiency in Buildings initiative in March 2006. United Technologies Corporation[10] and construction giant Lafarge are the two primary corporate partners, and the initiative aims to determine how buildings can be designed, constructed, and operated so that they use zero net energy, are carbon neutral, and can be built and operated at fair market value by 2050.[11] Initial findings indicate that technologies are available today to reach the energy and carbon goals. Also at the international level, the United Nations Environment Program (UNEP) announced in February of 2006 a Sustainable Building and Construction Initiative (SBCI). The purpose of the SBCI is to achieve worldwide adoption of sustainable building and construction practices that can address the issue of climate change.[12]

Finally, many corporations are taking progressive steps with the buildings that they construct, own, and operate. Wal-Mart, who has taken on aggressive targets to eliminate 30 percent of the energy used by its stores in the long term, is investing $500 million per year in technologies and innovations to reduce GHG emissions in stores around the world by 20 percent over the next seven years. It is also designing a prototype store that is 25–30 percent more efficient, produces 30 percent less emissions, and will be in operation within four years.[13] A number of other large corporations have taken on significant GHG reduction targets and have focused much of their efforts on their buildings. Swiss Re plans to be carbon neutral by 2013, and to achieve that goal, the company will have to reduce the emissions from its buildings by 33 percent.[14] Bank of America is in the process of building the first LEED Platinum[15] high rise as its new corporate headquarters; Toyota's headquarters in Torrance, CA is LEED certified; and Exelon Corporation is seeking LEED Platinum certification for its newly renovated headquarters building in Chicago.

## LOOKING AHEAD

Although private investment in green buildings and energy-efficient technologies is growing rapidly, coherent national policies on buildings are essential to address the built environment's role in climate change. The U.S. needs policies such as model building codes to raise the minimum standards for energy and GHG performance, as well as incentives for industry leaders to continually improve. One key focus should be on the construction of net-zero energy homes. On-site renewable energy such as solar photovoltaic technologies offer the possibility of net-zero-energy homes, when combined with 60–70 percent whole building energy reductions. This goal may be achievable as a cost-competitive housing alternative by 2020 (see Figure 4). The estimated cost premium for such a system today is approximately 25 percent.[16]

The U.S. Department of Energy's Building America Program is a public/private partnership that develops energy solutions for new and existing homes that can be implemented on a production basis (http://www.eere.energy.gov/buildings/building_america/). The program focuses on increasing the energy efficiency of homes and installing on-site renewable energy systems. Fully funding and expanding existing public/private partnerships such as the Building America Program and developing new, innovative partnerships to promote the growth of sustainable building practices at a commercial scale is essential to demonstrate the feasibility of these practices and the direct costs savings from decreasing energy use within homes.

The U.S. Department of Defense (DoD) can be used as a model for its work on net-zero energy buildings. Because DoD operates so many buildings in remote locations, it has invested substantial resources into funding remote power generation. In addition, DoD puts a lot of time, money, and research into developing the most efficient ways to use energy in its buildings. Domestically, DoD is the single largest energy user in the nation; it spends over $2.5 billion per year on facility energy consumption. Photovoltaic (PV) companies such as Daystar technologies and Evergreen Solar have won large contracts with DoD to develop affordable, renewable energy systems for buildings. By reducing the demand for energy within its buildings and increasing the supply of on-site, renewable energy, DoD is pursuing the goal of net-zero energy buildings, a perfect fit for its remote facilities. It is making a concerted effort to reduce the energy demand of its buildings

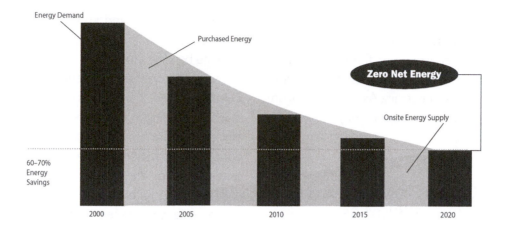

FIGURE 4. THE PATHWAY TO NET ZERO ENERGY HOMES.

in order to decrease its single largest operating cost.

## TREMENDOUS POTENTIAL FOR FUTURE PROGRESS

Applying currently available cost-saving and low-cost building design strategies, developed in the late 1970's and 80's, can cut fossil-fuel energy consumption in buildings by 30 to 80 percent. These include building siting, shape, color and orientation, and daylighting, passive solar heating, cooling, natural ventilation, and shading strategies.

New technologies can cost-effectively save an additional 30 to 40 percent of energy use and GHG emissions in new buildings, when evaluated on a life-cycle basis.[17] Technology opportunities are more limited for the existing building stock, and the implementation rate depends on the replacement cycles for building equipment and components. However, several opportunities worth noting apply to existing as well as new buildings, including efficiencies in roofing, lighting, home heating and cooling, and appliances. Emerging building technologies, especially new lighting systems and integrated

thermal and power systems, could lead to further cost-effective energy savings. Past experience has shown that policy intervention is most likely needed for serious market penetration of efficient energy systems and on-site electricity technologies.

In addition to building design strategies and technological advances in buildings, it is essential to focus on community and urban systems to cut associated GHG emissions from the building sector. Evidence suggests that higher-density, more spatially compact and mixed-use building developments can offer significant reductions in GHG emissions through three complementary effects: (1) reduced vehicle miles of travel, (2) reduced consumption for space conditioning as a result of district and integrated energy systems[18], and (3) reduced municipal infrastructure requirements. In total, therefore, smart land-use planning policies across the country could yield GHG reductions of 3 to 8 percent by mid-century.

Numerous individual, corporate, community, and state initiatives are leading the implementation of green building practices in new residential development and commercial construction. Significant progress in reducing GHG emissions from the building sector depends on federal, state, and local policymakers adopting whole building standards that put an emphasis on reducing these emissions. At the least, updating building codes to reflect the best local climate-specific codes is necessary. As shown in Figure 5, there is substantial opportunity to improve building codes.

## POLICY OPTIONS TOWARD ZERO NET-ENERGY BUILDINGS

Research suggests that public interventions could overcome many of the market failures and barriers hindering widespread penetration of climate-friendly technologies and practices. The mosaic of current policies affecting the building sector is complex and dynamic, ranging from local, state, and regional initiatives, to a diverse portfolio of federal initiatives. Numerous policy innovations could be added to this mix, and many are being tried at the state and local level.

Ten states[19] have set minimum energy efficiency standards for household and/or commercial appliances not covered by mandatory federal standards (see Figure 6). Without a waiver from the U.S. Department of Energy, states may not set standards for products covered by existing federal standards. Increasingly, many corporations are pushing for comprehensive federal standards that are applicable throughout the nation so they have a well-defined market against which they can judge their investment decisions.

Emissions can be addressed through labeling and expanded and tightened standards for products (including buildings), focusing on those that would result in significant GHG reductions through reduced energy use.[20] By requiring a minimal level of efficiency and providing consumers with information on products that do better than the minimum, standards and labeling can overcome the obstacles described earlier—insufficient and imperfect information; market distortions; and split incentives—and advance building efficiency.

According to the Pew Center's "Agenda for Climate Action,"[21] building codes can require that new buildings meet a certain level of energy efficiency, maximizing efficiency opportunities during construction. Policies to encourage states to adopt enhanced or updated building codes could include linking a state's adoption of model codes to its receipt of federal funds (e.g., weatherization assistance and federal support for state public benefit funds). Incentives could come in the form of a minimum requirement to receive federal funding (i.e., states would be required to adopt a certain standard level to be eligible for any funding), or as encouragement to receive additional or "bonus" funding (i.e., above the level that a non-adopting state receives). Increasing the funding level for the DOE's building energy code program would also facilitate GHG emissions reductions from further building code adoption—by providing stakeholders with energy efficient appliances, insulation in homes, and solar tax technical assistance such as software tools to help builders, designers, and code officials upgrade and comply with energy codes.[22] Likewise, continued funding for R&D on advanced materials and cost-reduction opportunities for on-site renewable generation can have a considerable impact.

The 2005 Pew Center report[23] reviewed buildings energy research and development (R&D) and six deployment policies that have a documented track record of delivering cost-effective GHG reductions and that hold promise for continuing to transform markets. The six deployment policies include (1) state and local building codes, (2) federal appliance and equipment efficiency standards, (3) utility-based financial incentive and public benefits programs, (4) the low-income Weatherization Assistance

Program, (5) the ENERGY STAR Program, and (6) the Federal Energy Management Program.

Annual savings over the past several years from these R&D and six deployment policies are estimated to be approximately 3.4 quadrillion Btu (quads) and 65 million metric tons of carbon (MMTC), representing 10 percent of U.S. $CO_2$ emissions from buildings in 2002. The largest contributors are appliance standards and the ENERGY STAR Program. Potential annual effects in the 2020 to 2025 time frame are 12 quads saved and 200 MMTC avoided, representing 23 percent of the forecasted energy consumption and carbon emissions of buildings in the United States by 2025.

While some of these policies were incorporated in the Energy Policy Act of 2005 (EPACT 2005), it is necessary to expand them further to realize more reductions in energy costs and GHG emissions from the buildings sector.

Several portions of the Energy Policy Act of 2005 focus on increasing building efficiency such as renewed incentives for credits. Focusing mainly on commercial products, the Energy Policy Act requires DOE to set standards for certain equipment and appliances including exit signs, traffic signals, torchiere lights, compact fluorescent lightbulbs, many types and sizes of heating and cooling equipment, refrigerators, freezers, automatic ice makers, clothes washers, and even spray valves. The Act also established tax credits for the construction of a qualified new energy-efficient home that meets Energy Star criteria and a tax deduction for energy-efficient commercial buildings that reduce annual energy and power consumption by 50 percent compared to the mandated standard. But it is essential to take steps beyond the Act to achieve the necessary GHG reductions in the building sector—for example, expansion of the Building America Program, a greater focus

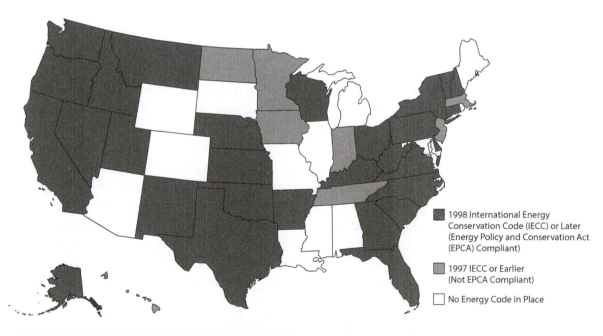

FIGURE 5. STATES WITH RESIDENTIAL ENERGY CODES.

FIGURE 6. STATES WITH ENERGY EFFICIENCY STANDARDS FOR HOUSEHOLD AND COMMERCIAL APPLIANCES.

on net zero energy buildings, and continuing to expand federal tax credits for on-site renewable power.

## CONCLUSIONS AND RECOMMENDATIONS

An expansive view of the building sector is needed to completely identify and exploit the full range of GHG-reduction opportunities. Such a view needs to consider future building construction (including life-cycle aspects of buildings materials, design, and demolition), use (including on-site power generation and its interface with the electric grid), and location (in terms of urban densities and access to employment and services).

There is no silver bullet technology in the building sector because there are so many different energy end uses and GHG-relevant features. Hence, a vision for the building sector must be seen as a broad effort across a range of technologies and purposes.

Furthermore, an integrated approach is needed to address GHG emissions from the U.S. building sector—one that coordinates across technical and policy solutions, integrates engineering approaches with architectural design, considers design decisions within the realities of building operation, integrates green building with smart-growth concepts, and takes into account the numerous decision-makers within the fragmented building industry.

Current building practices seriously lag best practices. Codes must be upgraded to improve the performance of the average building, and vigorous market transformation and deployment programs are critical to success. They are also necessary to ensure that the next generation of low-GHG innovations is rapidly and extensively adopted. To that end, green building standards must more explicitly incorporate GHG performance standards and metrics. In order for a building to be considered "green," it needs to fully address climate concerns.

Given the durable nature of buildings, the potential for GHG reductions resides both with new construction and the existing building stock for some time to come. The WBCSD initiative described earlier intends to commercialize newly constructed and retrofitted zero net energy homes and climate-friendly designs for large commercial buildings and industrial facilities by 2012. By mid-century, land-use policies could have significant impact on GHG emissions, as well. This inter-temporal phasing of impacts does not mean that retrofit, new construction, and land-use policies should be staged; to achieve significant GHG reductions by 2050, all three types of policies must be strengthened as soon as politically feasible.

Similarly, applied R&D will lead to GHG reductions in the short run, while in the long run basic research will produce new, ultra-low GHG technologies. This does not mean that basic research should be delayed while applied R&D opportunities are exploited. The pipeline of technology options must be continuously replenished by an ongoing program of both applied and basic research and a combination of public policies and private initiatives must pull these technologies into the marketplace.

## NOTES

1. This In-Brief draws heavily on the Pew Center report entitled: "Towards a Climate-Friendly Built Environment." Brown, M., Southworth, F., and Stovall, T. 2005. *Towards a Climate-Friendly Built Environment.* (Arlington, VA: Pew Center on Global Climate Change).

2. Carbon intensity is the ratio of carbon emissions to economic activity.

3. Building stock refers to the number of total square feet included in all building types.

4. This includes the electrical energy for industrial buildings operations.

5. A plug load is an appliance or a piece of household equipment that draws electrical power.

6. Under LEED, building projects are awarded points in six categories: sustainable sites, water efficiency, energy and atmosphere, incorporation of local and recycled materials and resources, indoor environmental quality, and innovation and design process. It has proved to be an effective voluntary standard, although some concerns exist regarding a lack of direct correlation between some of the points awarded and the life-cycle GHG reductions (or life-cycle costs) from the building.

7. See: Green Building Alliance. 2004. *LEED-NC: The First Five Years, Report on the Greater Pittsburgh Region's Experiences using Leadership in Energy & Environmental Design for New Construction,* Green Buildings Council, Pittsburgh, PA (www.gbapgh.org/MiscFiles/LEEDSurveyReport_Final.pdf. February 13, 2005). Also see: Rachel Reiss and Jay Stein. 2004. *LEED Scores Early Successes but Faces Big Challenges,* ER-04-3, Platts Research & Consulting, New York, NY.

8. California, Arizona, New Mexico, Colorado, Nevada, Washington, Wisconsin, Michigan, Connecticut, Rhode Island, and Maine require LEED certification for state and state-funded buildings, and New York, Pennsylvania, and New Jersey recommend LEED certification for state and state-funded buildings.

9. Carbon neutral means they will have zero net GHG emissions—either by using only non-fossil fuel energy or by offsetting their GHG emissions.

10. United Technologies Corporation provides a range of high-technology products and services to customers in the aerospace and buildings industry worldwide.

11. "Top global companies join with WBCSD to make energy self-sufficient buildings a reality." See: http://www.wbcsd.org/plugins/DocSearch/details.asp?type=DocDet&ObjectId=MTg2MTU, viewed 9/15/06.

12. "UNEP launches Green Building Initiative." http://www.unep.org/Documents.Multilingual/Default.asp?DocumentID=469&ArticleID=5204&l=en, viewed 9/15/06.

13. Scott, L. 2005. *21st Century Leadership*. Speech to company employees, October 24 (Bentonville, AR: Wal-Mart).

14. Hoffman, Andrew. 2006. *Corporate Strategies That Address Climate Change*. (Arlington, VA: Pew Center on Global Climate Change).

15. LEED Platinum is the highest ranking for certification under the U.S. Green Building Council's LEED program.

16. Zero Energy Homes brochure, NAHB Research Center, Upper Marlboro, MD, http://www.toolbase.org/docs/MainNav/Energy/4339_ZEH_Brochure-final-screen. pdf, December 6, 2004.

17. Brown, M., Southworth, F., and Stovall, T. 2005. *Towards a Climate-Friendly Built Environment*. (Arlington, VA: Pew Center on Global Climate Change).

18. District and integrated energy systems refer to systems that provide energy services (heating, cooler, electricity) to a community.

19. Arizona, California, Connecticut, Maryland, Massachusetts, New Jersey, New York, Oregon, Rhode Island, and Washington.

20. Examples include boilers and furnaces, digital cable and satellite TV boxes, and digital converter TV boxes. Nadel, S. 2003. "Appliance and Equipment Efficiency Standards in the U.S.: Accomplishments, Next Steps and Lessons Learned," ECEEE 2002 Summer Study Proceedings, European Council for an Energy Efficient Economy, 1: 75–86.

21. *The Agenda for Climate Action*. (Arlington, VA: Pew Center on Global Climate Change).

22. The Building Energy Code Program also provides financial and technical assistance to help states adopt, implement, and enforce building energy codes. For more information, see http://www.energycodes.gov/.

23. Brown, M., Southworth, F., and Stovall, T. 2005. *Towards a Climate-Friendly Built Environment*. (Arlington, VA: Pew Center on Global Climate Change).

Section VII

# COMMERCE AND PROTECTION

# THE BUSINESS CASE FOR THE GREEN ECONOMY

By United Nations Environment Programme

I n 2011 UNEP[1] produced a report titled *Towards a Green Economy*. It described an economy that results in improved human well-being as well as social equity, while significantly reducing environmental risks and ecological scarcities.[2] The report proposed a development pathway for policy makers to transition to a Green Economy that is low carbon, resource efficient and socially inclusive, and whose growth in income and employment is driven by public and private investments that reduce carbon emissions and pollution, enhance energy and resource efficiency, prevent the loss of biodiversity and ecosystem services and contribute to poverty alleviation. It demonstrates that a Green Economy will stimulate new markets, create more jobs, generate higher rates of GDP growth in the medium term, improve returns on investments, and reduce individual and collective risk, all whilst strengthening environmental, social and human capital.[3]

*The Business Case for a Green Economy*, complements and extends the 2011 report. Written for a corporate audience, it illuminates and clarifies the business benefits to companies that pro-actively participate in—perhaps even lead—the transition. Numerous examples and compelling empirical data demonstrate how business strategies that reflect the attributes of a resource efficient and green economy can positively impact the financial metrics of companies of all sizes.

These examples from both developed and developing countries, demonstrate how integrating sustainability into core business activities can generate a positive return on investment. Returns that go beyond the financial component and contribute to the socio-economic and environmental framework conditions necessary for business to grow and operate successfully.

An extensive body of research has been conducted to create this business case publication. Sources include UNEP's wider Green

Economy Initiative work, UNEP's partner organizations, and the wider business community. Numerous case studies included in this report reflect a new Green Business Case Model to demonstrate how actions taken by companies to improve their environmental impacts result in improvements to leading indicators of financial success, which result in improvements to six key financial metrics:[4]

- Sales growth
- Duration of sales
- Capital expenditure
- Profit margin
- Tax rates
- Cost of capital

As this report makes evident, the benefits to business of adopting greener and resource efficient practices are obvious, and the consequences to business of ecosystem collapse are disastrous. So, why are sustainability strategies not more widely adopted? Significant barriers remain, most notably the deep-seated financial short-termism that exists in businesses, markets and governments. In addition, many policies,

subsidies and incentives offered by governments and markets reward or encourage behaviours and decisions that increase, rather than decrease, environmental impacts. However, as pressure on ecosystems and natural resources increases, changes in public policy, customer preferences, and technology will drive the market for improved environmental performance. Businesses with the foresight to get in front of these changes will gain a competitive advantage. It is estimated that the annual financing demand required to create the Green Economy is in the US$ 1–2.5 trillion range.[5] This level of investment represents an enormous opportunity for the private sector to provide the infrastructure, equipment, goods and services that will drive the transition.[6] With this in mind, investors are increasingly considering environmental performance as a proxy for management quality.

The transition to a Green Economy is not an easy path. It is characterized by step changes in resource efficiency and a shift in emphasis from shareholder value to stakeholder value. Some companies, and perhaps whole industries, will not survive the transition. Success over the long-term will require new skills, diverse

---

### The Business Case for the Green Economy at a Glance

Decoupling environmental impact enables companies to position themselves for sustainable business growth. The Green Business Case Model identifies actions companies can take to drive leading indicators of performance and improve financial value drivers.[6]

### Green Economy benefits to business include:

- More resilient supply chains
- New investment opportunities
- Increased consumer demand for sustainable goods and services
- Sales growth and duration of sales
- Training and job creation
- Reduced dependency on natural resources
- Mitigation against the negative financial risk from environmental impact

collaborations, continuous innovation, investments with uncertain returns, and a change in what the market values. Companies, like governments, will need to choose wisely if they are to capitalize on the opportunities it brings.

The gathering of leaders from government, business and civil society at the Rio+20 United Nations Conference on Sustainable Development presents a historic opportunity to accelerate the transition to a Green Economy. Positive business engagement in the conference is critical. Indeed, while public policy is an essential ingredient in making the Green Economy a reality, it is the actions of the private sector that will ultimately determine whether and how quickly it occurs.

## Business Actions:

- *Enhance resilience and business growth by adopting alternative valuation techniques:* Traditional monetary valuation techniques fail to capture the value of supplies from nature. Alternative techniques more adequately value human, social and natural capital.
- *Drive policy change:* Companies can also drive policy change for market-based regulations of environmental "bads" and cuts in public funding in sectors that deplete natural capital. This can create opportunities for responsible business investing in the technology innovations and entrepreneurship that these new markets require.
- *Move from shareholder to stakeholder value considering the wider operating framework:* Businesses are achieving competitive advantage by looking at their company as part of a wider network of stakeholders and engaging them to make better informed decisions.

- *Ensure employee engagement and enhance resource productivity:* Creating incentives and mechanisms to embed sustainability within company culture in all operations will reap more societal benefits and resource efficient outcomes.
- *Establish sustainability as a permanent item on the Board agenda and communicate its value to investors and consumers:* By placing sustainability at the core of governance, leading companies are planning for the implications of the transition to a Green Economy. Making the link between sustainability and financial reporting will enable the value to be better communicated to investors and consumers.

As companies anticipate and drive the transition to a Green Economy, the business case will grow. So the question is: why wait?

## INTRODUCING THE GREEN ECONOMY AND BUSINESS

The Green Economy provides a clear opportunity to boost economic development at a time of low GOP growth and recession. Conventional methods to promote economic recovery are becoming more limited and therefore business and governments are seeking new ways to create long-term prosperity in a resource-constrained world. This transition requires a shift away from business-as-usual.

## Drivers of Change

### Resource Pressure

Driven by population growth and additional consumer pressure in emerging economies, global demand across all major environmental resources will increase over the coming years.

Between 2010 and 2030, demand is projected to grow by 33 percent for primary energy, by 80 percent for steel,[7] by 27 percent for food (cereals) and by 41 percent for water (see *Figure 1*).[8] This combined with risks and environmental stresses derived from climate change, such as water and land availability and biodiversity loss, means that the pressures on these resources are considerable and likely to grow.

## Constraints in Production

Environmental factors are acting as constraints in production, particularly in the agricultural and energy sectors, heavily dependant on these resources. This confirms that business as usual over the long-term is simply unviable. Companies acting ahead of the game will be able to tap into most of the opportunities available, sustain growth and adjust to changes to the regulatory environment which, in turn, is responding to the growing evidence of environmental degradation.

Take the manufacturing sector for example. The sector is responsible for 35 percent of electricity use, more than 20 percent of $CO_2$ emissions and over a quarter of primary resource extraction, directly impacting economic growth, the environment and human health.[9] By implementing sustainability measures the manufacturing sector can boost economic and environmental performance. Improving recovery and recycling, the adoption of closed-cycle manufacturing and extending the lifespan of manufactured goods can help to decouple this sector's growth from its environmental impact. Examples of this are given throughout the report.

Remanufacturing operations worldwide already save 10.7 million barrels of oil each year. Sustainable production practices can reduce emissions and integrate by-products into the production value chain. As such, return on investment can be substantial and payback periods reduced. There are also positive

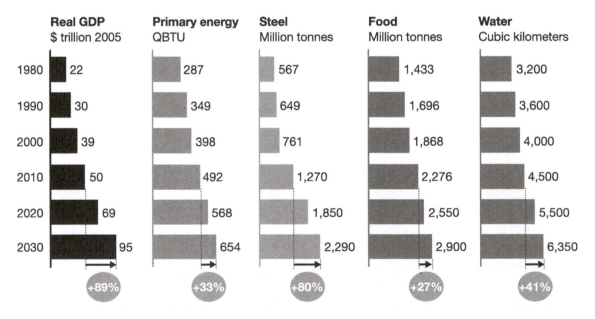

FIGURE 1. HISTORICAL AND ESTIMATED GROWTH IN DEMAND FOR RESOURCES FROM 1980 TO 2030.

Source: McKinsey & Co Global Institute (2011). Resource Revolution: Meeting the world's energy, materials, food, and water needs, p. 35. New York. Available at : http://www.mckinsey.com/Features/Resource_revolution.

implications for jobs through opportunities in secondary production.

### Consumer Demand

Many companies see sustainability as a key driver of innovation and new sales opportunities. Increasingly well-informed consumers are raising their expectations of the private and public sectors. Consumer demand for sustainable products is growing, and there is much untapped latent demand that can be unlocked by companies with a clear vision of a more sustainable future. Leading companies are using sustainability as a strategic driver, informing structural changes, and governance and investment decisions. However, the majority of businesses require more persuasion.

## Critical Choices

Critical choices need to be made now, to decouple economic growth from environmental impact and resource use, to prepare for the economic reality of tomorrow, and to enable business to seize the scale of the opportunity. To create resilient systems, substantial innovation is required. A more resource efficient and Green Economy provides the framework for a stronger and more sustainable business approach, at the heart of which lies the necessity to ensure sustained financial growth over the long-term.

A Green Economy requires step changes in resource efficiency, investment in clean technologies, the development of alternative products, services and materials, and the ability to obtain value from unavoidable waste. *Figure 2* shows the key material issues and summary implications for a number of selected sectors. Those at the forefront of this evolution are developing new governance models, enhancing management processes, and developing measurement and reporting tools that can more adequately account for the complexities of sustainability.

## Collaborative Capitalism

Brand value benefits associated with sustainability are influencing business-to-business relationships and opportunities. Corporate customers with active sustainability strategies are placing significant demands on suppliers to meet new standards; for example, on product environmental performance and respecting human rights along the supply chain. We are entering an age of collaborative capitalism, where achieving business objectives are dependent on the creation of shared and common objectives. Stakeholder engagement, life-cycle thinking and value chain management all create the potential to collaborate with a common purpose—increasing resource efficiency and social cohesion.

The current business environment is one of intense competition, pressure on operating margins, and, in some high-profile cases, bankruptcies and business failure. This leads to mixed messages from policy makers and regulators. In efforts to support traditional industries, there are many examples of governments scrapping or scaling back regulatory incentives for sustainable innovation. However, many, including the World Business Council for Sustainable Development (WBCSD), recognize that the introduction of regulation and market incentives at scale can stimulate cleaner economies—the so-called 'Green Race'.[10] In the past decade, carbon markets have been developed in many countries and have helped to stimulate innovation and efficiency. The valuation and regulation of broader eco-system services may well follow this path.

## The Value Proposition

Pioneering companies with vision create new markets and see the potential where others see risk and liability. Increasing consumer awareness drives demand for sustainable products,

FIGURE 2. MATERIAL USES AND IMPLICATIONS BY SELECTED SECTORS.

Source: *Two Tomorrows* (2012). London.

| Sector | Material Issues | Implications for transition to Green Economy |
|---|---|---|
| **Building and construction** Housing; industrial and infrastructure construction | • Energy use and emissions<br>• Materials use<br>• Waste<br>• Water<br>• Health and safety | • Resource efficiency<br>• Sustainable innovation<br>• Life-cycle management |
| **Food and Beverage** Agricultural production; processing; food production; distribution; retail; and catering | • Overfishing<br>• Increasing meat consumption<br>• Environmental degradation<br>• Health<br>• Resource and energy use<br>• Water<br>• Waste<br>• Worker rights | • Rebuild overfished and depleted fish stocks<br>• Adopt resource-conserving practices<br>• Focus on health and well-being<br>• Improve supply chain conditions |
| **Transport** Air; rail; shipping; road vehicles | • Environmental degradation and land use<br>• Alternative fuels<br>• Road safety<br>• Emissions | • Develop lighter vehicles<br>• Increase emphasis on collective and intermodal travel<br>• Increase use of technology to reduce impacts |
| **Tourism** Lodging; recreation; restaurants and bars; events; tourism services; culture; tours and excursions | • Environmental degradation<br>• Socio-economic development<br>• Heritage and culture<br>• Water and resource use<br>• Waste<br>• Human rights | • Collectively manage resources between private and public partners<br>• Improved valuation of heritage and culture |
| **Extractives** Mining of minerals and metals; oil and gas | • Environmental degradation<br>• Energy and water use<br>• Human rights<br>• Host community benefits including employment, revenues and taxes<br>• Health and safety | • Prepare for a shift in the pricing structure of resources<br>• Increase energy efficiencies<br>• Life-cycle resource management |
| **Utilities** Water and waste management; energy production and distribution | • Emissions<br>• Nuclear power<br>• Water use<br>• Waste disposal | • Manage demand<br>• Increase efficiencies<br>• Develop renewable technologies at all scales<br>• A shift in the pricing structure of resources<br>• Waste to resource – reuse, recycling and energy recovery from waste |

and globally expanding businesses see that this is particularly true of emerging markets. A National Geographic/GlobeScan 2010 survey of 17,000 people in 17 countries found that consumers in Brazil, India and China scored the highest in terms of increasing environmentally sustainable consumer behavior.[11] This is based on factors such as the energy and resources consumed per household and changes within the categories of personal transportation, food and consumer goods. The new lifestyle markets, markets for sustainable cities, the service markets, the organics and certified markets for aware consumers, are all examples of opportunities to be cultivated and seized.

This value proposition remains strong for all scales of enterprise including Small and Medium-sized Enterprises (SMEs). More sustainable products and services can help SMEs drive up quality and financial and socio-economic benefits. The demand is growing from a larger customer base—3 billion middle class consumers alone are anticipated in the next decades—and changing public-sector expectations. Collaboration with them will help facilitate this change, building economies of scale.

## LEVERAGING BENEFITS THROUGH THE GREEN ECONOMY

There is a compelling case for the transition towards a resource efficient and Green Economy. Decoupling is good for business: rather focus only on cost, an equally pertinent question is, 'to what extent do these actions have a positive impact?' The challenge is how business can justify sustainable decision-making.

## The Green Business Case Model

The Green Business Case Model, as shown by *Figure 3* [page 240], demonstrates the correlation between the greening of business and financial value creation and offers a framework to support this process. It highlights actions that improve business resilience, deliver enhanced financial performance, and, at the same time, support broader environmental and socio-economic goals.

The model suggests a circular chain of cause and effect. It has three components:

1. Taking action
2. Leading indicators
3. Financial value drivers

## Applying the Model

Society has so far failed to translate the clear macro-economic case for sustainable development to a business case at enterprise level. The evidence throughout this report provides examples of how the model is being applied across a range of sectors, geographies and scales of enterprise. Though the business case is sound for all sizes of business, it may be useful to consider it across the value chain and at various scales.

The model highlights a virtual circle driven by the need for sales growth and duration; improved profit margin and reduced capital expenditure; and preferable tax and reduced cost of capital. All of these can be enhanced by taking sustainable actions that improve the leading indicators of business success. ...

## SALES GROWTH AND DURATION

As market and regulatory demand for sustainability grows, the business that makes effective use of tools such as **design for sustainability**

FIGURE 3. GREEN BUSINESS CASE MODEL.

and the delivery of more **sustainable goods and services** will be in a position to boost its ability to innovate and to **attract more customers**, which in turn will show positive results through **sales growth and duration**.[12]

## Attracting More Customers

Consumer interest in sustainable products is progressing from early fears about assumed higher cost to a trend in which positive environmental and social elements are an inherent part of product quality. Recent survey results[13] suggest that in future business-to-business (B2B) and business-to-consumer (B2C) customers and other stakeholders will expect all products to be environmentally and socially responsible. Demand has generally been resilient with many customers willing to pay a premium for sustainability credentials across most categories.

Attracting more customers depends on the effective promotion of the benefits of sustainable consumption, which can include:[14]

a. **eco-innovation** and **design for sustainability** to deliver maximum societal value at minimum environmental cost
b. choice influencing through marketing communications and awareness-raising campaigns
c. choice editing by removing unsustainable options from the market place.

For example, product service models enable a move beyond business-as-usual approaches that encourage the quick disposal of products or "rapid obsolescence" and its related waste of resources. Business for Social Responsibility (BSR) argues that sustainable consumption can drive and define innovation in the world's fastest growing markets, noting that nearly

80 million people are joining the middle class in emerging markets every single year.[15] This shows the clear need to rethink how economic growth is pursued and business models applied. The reorientation of consumer choices and lifestyles will play a crucial role.[16]

Information across all markets and sectors is not always available, but most surveys and data indicate **sales growth**. For example, major markets for organic food and beverages have expanded by 10–20 per cent in the past decade.[17] When done accurately and well, to robust standards, labels and certification can improve access to more reliable information and help to inform consumers and boost the uptake of more **sustainable goods and services**.[18]

## New Markets

Considerable new market opportunities lie in developing sustainable business models that help improve the lives and environments of low-income consumers. Four billion people at the base of the pyramid, with annual incomes below US$ 3,000 in local purchasing power, live in relative poverty yet represent a market of US$ 5 trillion. Combined with the growing affluence of the new middle classes, emerging economies represent key market growth opportunities. *Unilever*, for example, expects 70 per cent of its business to come from Asia within 10 years.[19]

Environmental policies and regulations are also creating new commercial opportunities, for instance through carbon and ecosystem service markets. Markets for biodiversity offsets are predicted to grow to US$ 10 billion by 2020, and global revenues for companies involved in the wind, solar and biofuels markets have seen a ten-fold rise to US$ 116 billion over the five years to 2008. Despite the setback in 2009, due to the global recession, the long-term prognosis is still for a rapid rise, to more than US$ 300 billion annually by 2020.[20] As the *Siemens* case

study illustrates, more efficient **clean technologies** at scale are increasingly competitive and driving **sales growth and duration**. ...

Systems and life-cycle thinking also drive new market development such as the emergence of vertical integration along supply chains. To this end, a number of manufacturing firms are moving into project development, retailing, installation, product take-back and after-sales services.

The Integrated Multi-Trophic Aquaculture (IMTA) approach employed by China's *Zhangzidao Fishery Group* offers an alternative to monoculture methods, providing a more balanced ecosystem that takes into account local conditions, operational limits and environmental quality. This enabled the company to see revenue grow annually by 40 per cent between 2005 and 2010, compared to the industry average of 13 per cent. Its average EBITDA margin was 31 per cent.[21]

## Building Brand Value

Businesses that actively promote more **sustainable goods and services,** backed up by consistently recognized standards and labels, will benefit from enhanced **brand value and reputation.** This enables them to sustain **growth of sales** with longer **duration,** reflecting greater loyalty among existing customers and continual improvement in reaching new customer bases.[22]

*Unilever*, for example, with its Sustainable Living Plan is actively promoting more **sustainable goods and services** with time bound public targets and progress reports. This action has a high potential for increasing its brand value.

There is evidence of a correlation between a company's visible commitment to environmental and socio-economic commitment and principles, and the value of that company's corporate and product brands.[23] Since **brand value** is vulnerable to reputational risks, including, for instance, human rights, bribery and corruption or environmental incidents, transitioning to a resource efficient and Green Economy[24] is expected to have a positive impact on **brand value.**

The global digital revolution has improved access to information regarding companies and their brands and this has the potential to make or break a company's reputation. Poor reputations take a long time to fix and good reputations are strongly linked to resilience and enduring organizational success. *Figure 5* highlights how brand reputation is often seen as the most important reason for organizational response to sustainability.

## REDUCED CAPITAL EXPENDITURE AND IMPROVED PROFIT MARGIN

More sustainable resource use, and more efficient supply chains, are increasing operating and net profit margins, with businesses earning more per dollar of sales thanks to lower production costs and **reductions in capital expenditure.** For example, improved **resource efficiency** is driving better use of fixed assets such as land, buildings, equipment and vehicles. In the case of working capital, efficiency improvements serve as a driver for innovation in the way inventory and customer and supplier relations are managed.

### Resource Productivity

Operational **resource efficiency** activities have been widely adopted—though far below their potential—as they generally 'pay in' using traditional cost/benefit analysis. However, due to the scale of change required if companies are to operate within a Green Economy, there is still enormous scope for further improvements by building on **efficiency** (doing the same with less) to **resource productivity** (doing better with

less). While existing literature demonstrates that massive improvements in **resource efficiency** exist at the process level, many companies are rethinking the way they do business from a systems perspective, leading to **resource productivity** improvements. While any enterprise can benefit, this is especially critical to industries in very competitive markets.[25]

Take the example of *General Motors*, which implemented its Resource Management Programme at many of its manufacturing facilities. Through improvements in **resource productivity,** the company has saved more than US$ 30 million and reduced waste volume by 40 per cent from 2000-2008. Furthermore, *General Motors'* efforts to transform waste into valued by-products created US$ 6 million in sales.[26]

Businesses can improve their **efficiency** by adhering to recognized standards and adopting cleaner technologies, both of which are conducive to more sustainable resource use. They also improve efficiency through **resource productivity** which involves re-thinking systems to maximise value-added of the resources used. An example is to move from selling a product to a product-service system, which can lead to more resource efficient and durable designs amenable to repair and recycling. Chemical leasing is one such service-oriented business model, where there is a shift from the sale of chemicals to the sale of functions performed by the chemicals, thus reducing inputs and waste outputs. There are increasing examples available in technology and systems options for energy, water and supply chain efficiency and waste reduction measures as well as new innovative resource **efficiency** and **productivity** options, through which businesses will ultimately develop a structural cost advantage.[27]

The expansion in scope of the Greenhouse Gas (GHG) Protocol to products and indirect 'Scope 3' emissions illustrates how collaborative

FIGURE 5. SUSTAINABILITY BENEFITS—BRAND REPUTATION AT THE TOP IN A SURVEY.

Source: Management Review and The Boston Consulting Group (2011). Sustainability: The "Embracers" Seize Advantage. Boston: Massachusetts Institute of Technology and BCG.

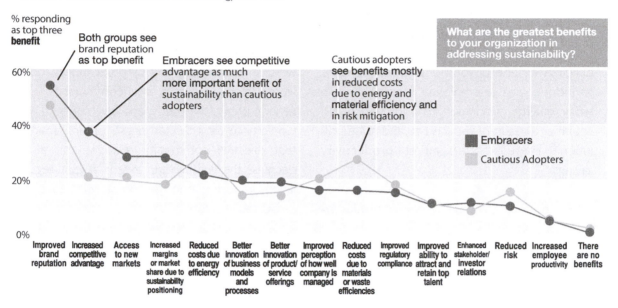

tools are enabling supply and value chain impacts to be understood and managed. These tools take a life-cycle approach to greenhouse gas accounting, which can draw attention to upstream or downstream opportunities. For instance, *PepsiCo* discovered extensive greenhouse gas emissions associated with the fertilizer applied to the oranges grown for its Tropicana juice. As a result, it worked with its orange growers to adapt the fertilizers used thereby reducing emissions.

Innovative accounting practices can help identify and focus **productivity** opportunities. … *PUMA*, the German sports apparel and footwear company, which has developed an environmental profit and loss account (EP&L) to quantify the value of ecosystem services and impacts. The EP&L provides insight into strategic environmental risks across its operations and supply chain, helping to focus **productivity** efforts where they can be most effective.

## Productivity (Including Human Capital)

A recognized challenge when assessing sustainable return on investment is being able to understand and demonstrate the links between greater care for human and social capital and environmental responsibility. Synchronous adoption of higher environmental standards with improved human resource management can be achieved through employee engagement programmes with an inherent link to sustainability performance, in addition to reward and incentive schemes.[28]

When sustainability efforts are implemented effectively, businesses can expect to improve their attractiveness to recruit new and top talent, to retain employees and to enhance employee and supplier **productivity.** These drive longer-term benefits such as customer attraction, improved reputation, a boost in operating margin, as well as optimized capital expenditure.

Evidence suggests that even subtler well-being changes can deliver significant productivity gains. There are considerable benefits to investing in air quality, lighting and health. The Green Economy Report shows results from a series of studies on the effects of better environmental conditions within workplaces. The advantages of green or sustainable buildings extend beyond environmental gains, to include greater employee productivity and **work quality,** in addition to improved public health as a result of reduced air and noise pollution.[29] These indicate significant **productivity** savings:

- Indoor air quality: 6–9 per cent **productivity** gain
- Natural ventilation: 3–18 per cent **productivity** gain
- Local thermal control: 3.5–37 per cent **productivity** gain
- Daylighting: 3–40 per cent **productivity** and sales gain

A number of studies[30] conclude that business can improve wellness, prevent chronic diseases and reduced healthcare costs while reaping the dual benefits of improved financial results and heightened employee goodwill.[31]

## PREFERABLE TAXATION AND REDUCED COST OF CAPITAL

Regulators and capital providers are raising their expectations of business. For example, disclosure and reporting requirements against environmental and social metrics is expanding in scope and sophistication. For companies trailing behind in the race to a more resource efficient and Green Economy, these requirements imply greater burden and cost, but for the leaders, who take early action, the rewards are significant, facilitated by new preferable tax regimes and reduced cost of capital.

### Licence to Operate

Early adapters looking to manage their strategic and operational risks and opportunities are implementing procedures for systematic stakeholder engagement. Building trust in this way can help secure a 'licence to operate', which in effect is 'granted' by stakeholders such as regulators, politicians, local communities, the general public, the media and civil society. The Conference Board,[32] amongst others, found that effective community relations can also decrease the number and extent of local regulations imposed upon the company because of its trusted reputation. ...

### Tax Rates [33]

Policy makers have a key role in improving tax regimes so that they recognize and award sustainable innovations and encourage verifiable improvements in operating conditions. Companies have more leverage where previously externalized impacts are reduced. In Guatemala, for example, tax breaks are provided on equipment for project developers to support a goal of 60 per cent of electricity being generated from hydro and geothermal by 2022. And the town of Caledon, in Ontario, Canada, offers property development fee discounts of five per cent if projects include renewable.[34]

Rather than higher expectations and requirements leading to a 'race to the bottom' with companies fleeing to jurisdictions with the lowest enforcement of regulation, the business case review suggests a 'race to the top'. Locations with higher environmental standards are already becoming more dynamic and attractive for investment and tax subsidies in favour of environmentally preferable goods and services.

## Cost of Capital and Risk Profile

Environment, health, safety and other human well-being issues are starting to feature more predominantly in **risk management.** Businesses with effective and well-communicated environmental and social **risk management** systems in place are in a position to secure a better **risk profile,** thus enabling them to obtain capital at lower cost. This applies to both debt capital and equity capital, and average cost of capital. ...

Climate change is opening up new opportunities for sustainable products and services in the finance sector. For instance, there is a pressing need to increase availability of capital to sustainability entrepreneurs and to further develop insurance schemes against environmental risks. The case study shows how Equity Bank of Kenya has **grown revenues** and **profits** partly through providing loans at market-leading interest rates to farmers who introduce environmentally preferable practices such as drip irrigation and water efficiency projects. The bank also supports farmers' financial literacy and skills development, helping to economically empower them.

### Securing Preferable Investment

Companies of all sectors will increasingly see that improved environmental **risk management** enables them to also obtain equity capital at lower cost. A statistical analysis of companies in the United States listed on the Standard & Poor's 500 dataset has shown that firms that lower their systemic **risk profile** through improved environmental **risk management** experience less volatility in performance and are rewarded by lower costs of equity capital and ultimately lower weighted average cost of capital[35]

The *Yes Bank* of India promotes investment in cleaner technologies and its responsible equity investment programme, 'Tatva', includes special consideration for SMEs. In Japan, the *Bank of Tokyo-Mitsubishi UFJ* has introduced products including the ECO Accreditation Support Loan, Eco-action Business Loans and Eco Stage Bond with preferential rates and fees for SMEs. Banks such as the Dutch *Rabobank* have also moved into technology leasing, providing environmentally friendly technologies at preferential rates to commercial customers.

Moreover, financial institutions can themselves benefit from transitioning towards a Green Economy. Environmental and sustainability risks are increasingly impacting credit defaults and banks' loan portfolios. By addressing ESG risks, they place themselves in a better position to access finance from capital markets or from multilateral financial institutions. Banks that incorporate criteria related to environmental **risk management** benefit from increased security and enhance their competitiveness, as more and more businesses align to this approach.[36]

## CHALLENGES

The business case for a Green Economy is becoming clearer, so it is surprising that more businesses are not joining the race 'to the top.' In 2011 GlobeScan, SustainAbility and UNEP conducted a survey of sustainability experts and practitioners,[37] which showed the perception that policy is not aligned with stated political goals of sustainability.

## Financial Short-Termism

An overwhelming 88 per cent of respondents cited the long-standing problem of financial short-term ism as the most important barrier for developing sustainability focused business models. Resource scarcity is driving demand and pushing up prices. Some investors and businesses have sought to exploit these trends through speculation and hedging strategies designed to deliver quick returns rather than protecting the natural assets that business and society depend on.

Acceptable return on investment time frames vary by industry sector, for example from 12 months in the ICT sector to more than 30 years in the power-generation sector. These time frames often conflict with appropriate time frames for regeneration and recovery of ecosystems. Longer return on investment time frames are increasing, but this is not widespread.

## Inappropriate Regulation

Regulations that inhibit change, combined with a lack of regulation that encourages more sustainable practices, often undermines progress towards a Green Economy, with 65 per cent of survey respondents citing regulatory regimes as a key barrier. To overcome these limitations, businesses can encourage policymakers to adopt policies that support sustainable business practices and innovation.

Inappropriate regulation also means that market failures continue to disincentivize change. Where natural resources are not priced, or are mis-priced, few businesses have a financial case for responding sustainably. For example, water use subsidies in many agrarian regions mean that farmers and agri-businesses largely ignore growing water scarcity. The issue is even more severe when it comes to the costs of dwindling biodiversity stocks,[38] which are still largely invisible in markets. The business case will be strengthened if these market failures can be addressed.

## Lack of Understanding of Business Imperative

Similarly, 65 per cent of respondents to the survey indicated that low awareness of the sustainability imperative among business leaders was also a significant barrier. The survey results imply that if executives understood the risks and the opportunities that issues such as human rights, climate change and water scarcity represented to their businesses, the level of resource commitment—and consequently the pace of change—would be dramatically increased.

## Generating Sufficient Market Pull

More than half of the respondents indicated that despite increasing market and regulatory demand for sustainable products and services, the low overall base from which this demand is growing indicates that sustainable business practice is not yet mainstream.

## Lack of International Standards

Half of respondents pointed to a lack of international standards as being a barrier. While there has been tremendous growth in sustainability standards, these have often been voluntary and when legislated, the requirements are often poorly enforced. A number of international guidelines and standards, such as the Global Reporting Initiative (GRI) and the ISO 26000 Social Responsibility Standard, are becoming widely adopted by many of the world's largest enterprises. These leading businesses are now calling for mandatory social and environmental reporting in order to create a level playing field and drive up performance.

## Delivery

SMEs face a unique challenge regarding the business case for transitioning to a Green Economy. For them, economies of scale may not be achievable, so there is a need to raise awareness of the potential quality, financial and socio-economic benefits that can be gained through a shift to **resource efficiency.** With the majority of economic activity flowing through

SMEs and value chains, mobilizing all parts of the economy fully and quickly is imperative if a sustainable path is to be found.

## CONCLUSION

The transition toward a resource efficient and Green Economy clearly represents an enormous opportunity for business. Those companies that understand and act on that opportunity will have an advantage in the market. However, there is also overwhelming evidence in the Green Economy Report and a number of other research based reports that business as usual is fundamentally undermining the very resource base upon which business success depends. Decoupling economic growth from environmental damage is required to prevent large-scale economic as well as environmental disaster. The drivers of change such as resource scarcity, stakeholder demand, and strategic opportunity are all accelerating. Business as usual is not an option.

The evidence within this report and that garnered during the wider research review upon which this report was based unequivocally demonstrates that positive returns are being achieved on investing in a Green Economy. There is a compelling case that **sustainable actions** lead to improvements in **leading indicators** of success, which in turn enhance **financial value drivers**. Pioneers that are leading the market are reaping the rewards and positioning themselves for sustained success.

Innovating to create more sustainable products has been shown to drive **customer attraction**, increasing **sales growth** and market share. The provision of sustainable products and services bolsters **brand value** and **reputation**, which in turn increases **sales duration**. Improving **productivity** strengthens **profit margins**. Business will be rewarded for sustainable

actions through favorable tax regimes which can reduce cost of capital.

It is not always easy to construct and embed a business case for change in the current business climate as barriers remain. Financial short-termism, regulations that encourage unsustainable practices, and low market pull reinforce the status quo. Building a business case for an individual company to change course requires both vision and courage, and must consider the context of the company, its industry, location and resource issues.

Business alone cannot deliver the speed and scale of change required. Collaboration with regulators, customers and the financial community is essential. Public policies linked to clear principles of sustained economic success are necessary to support this transition. Financial institutions playa dual role in the transition towards a Green Economy, through both investing in sustainable projects and integrating environmental, social and governance (ESG) indicators into the decision-making criteria for lending, investment and insurance. ESG performance is increasingly seen as a proxy for management quality, which in part explains the growing interest in sustainability ratings schemes.

From utility companies in the burgeoning urban centres of tomorrow to consumer goods companies in developed markets, **sustainable goods and services** are moving from niche to mainstream. Leading businesses have recognized the risks of financial short-term ism and have developed forward-thinking strategies. Those investing in sustainable innovation to increase **resource efficiency** and responsible operations ahead of regulation are achieving competitive advantage and positioning themselves to capture the mainstream markets of the next decade. As market conditions change, the business case will grow. So the question to business is: why wait?

## NOTES

1 World Business Council for Sustainable Development (WBCSD) (2008). *Sustainable Consumption Facts and Trends—From a Business Perspective.* Geneva: WBCSD. Available at: http://www.wbcsd.org/pages/edocument/edocumentdetails.aspx?id=142&nosearchcontextkey=true.

2 United Nations Environment Programme (UNEP) (2011). *Towards a Green Economy: Pathways to Sustainable Development and Poverty Eradication.* Available at: http://www.unep.org/greeneconomy.

3 Ibid. See also: Organization for Economic Cooperation and Development (OECD) (2011). *Towards Green Growth. A Summary for Policy Makers.* Paris: OED.

4 These metrics are combined into three overall financial drivers in the Green Business Case Model: sales growth and duration; reduced capital expenditure and improved profit margin; and preferable taxation and reduced cost of capital.

5 UNEP (2011). *Towards a Green Economy: Pathways to Sustainable Development and Poverty Eradication.* Available at: http://www.unep.org/greeneconomy.

6 Please refer to page 10 for the Green Business Case Model.

7 Steel has been considered as proxy measure for resource consumption in the global economic context.

8 McKinsey & Co Global Institute (2011). *Resource Revolution: Meeting the world's energy, materials, food, and water needs,* ch. 2. New York. Available at: http://www.mckinsey.com/Features/Resource_revolution.

9 UNEP (2011). *Towards a Green Economy: Pathways to Sustainable Development and Poverty Eradication,* p. 241–286. Available at: http://www.unep.org/greeneconomy.

10 WBCSD (2010). *The Green Race Is On* [online]. Available at: http://www.wbcsd.org/Pages/EDocumenVEDocumentDetails.aspx?ID=12772&NoSearchContextKey=true.

11 National Geographical Society and Globescan (2010). *Consumer Choice and the Environment. A worldwide tracking survey* Toronto: GlobeScan and National Geographic Society.

12 These actions can include the range of sustainability tools and approaches that can assist in assessment, measurement, implementation and communication.

13 PricewaterhouseCoopers (PwC) (2011). *Sustainable Growth. 14th Annual Global CEO Survey.* Available at: http://www.pwc.co.uk/ceo-survey/whats-on-themind-of-communications-ceos.jhtml. See also: KPMG (2011). *International Survey of Corporate Responsibility Reporting 2011.* Amsterdam: KPMG.

14 WBCSD (2011). *A vision for sustainable consumption—Innovation, collaboration, and the management of choice.* Geneva: WBCSD.

15 Business for Social Responsibility (BSR) (2010). *The New Frontier in Sustainability—The Business Opportunity in Tackling Sustainable Consumption.* San Francisco: BSR. See also: World Economic Forum (WEF) and Accenture (2012). *More with Less: Scaling Sustainable Consultation and Resource Efficiency.* Geneva: WEF.

16 UNEP (2011). *Paving the way for Sustainable Consumption and Production: The Marrakech Process Progress Report.* Available at: http://www.unep.fr/scp/marrakech.

17 Based on the data from Sahota, A. (2009). The Global Market for Organic Food & Drink. In: Willer, H. and Kilcher, L. (eds.) *The World of Organic Agriculture: Statistics and Emerging Trends 2009.* FIBL-IFOAM Report.

18 Porter, M. and Kramer, M. (2011). Creating Shared Value. *Harvard Business Review* (January–February), p. 53-67.

19 International Finance Corporation (IFC) (2011). *Inclusive Business Models—Guide to the Inclusive*

Business Models in IFC's Portfolio: Client Case Studies. Washington, DC: IFC.

20  Bishop, J. (ed.) (2011). *The Economics of Ecosystems and Biodiversity in Business and Enterprise.* London: Earthscan.

21  23 UNEP (2011). Investing in Natural Capital: Agriculture. In: *Towards a Green Economy: Pathways to Sustainable Development and Poverty Eradication.* Available at: http://www.unep.org/greeneconomy.

22  WEF (2011). *Redefining the Future of Growth: The New Sustainability Champions.* Geneva: WEF.

23  Sustainability and UNEP (2001). *Buried Treasure: Uncovering the Business Case for Corporate Sustainable Development.* London: SustainAbility.

24  SustainAbility and UNEP (2001). *Buried Treasure: Uncovering the Business Case for Corporate Sustainable Development.* London: SustainAbility.

25  UNEP (2011) *Decoupling natural resource use from environmental impacts and economic growth,* A report of the International Resource Panel, Available at: http://www.unep.org/publications/.

26  WBCSD (2008). *Sustainable Consumption Facts and Trends—From a Business Perspective.* Geneva: WBCSD. Available at: http://www.wbcsd.org/pages/ edocumenUedocumentdetails.aspx?id= 142&nosearchcontextkey=true.

27  McKinsey & Co Global Institute. (2011). *Resource Revolution: Meeting the world's energy, materials, food, and water needs.* New York. Available at: http://www.mckinsey.com/Features/Resource_revolution.

28  Carbon Disclosure Project (2010). *Carbon Disclosure Project 2010—Global 500 Report.* UK: PwC.

29  UNEP (2011). *Towards a Green Economy: Pathways to Sustainable Development and Poverty Eradication.* Available at: http://www.unep.org/greeneconomy.

30  PwC and WEF (2008). *Working Towards Wellness: The Business Rationale.* Available at: http://www.pwc.com/gx/en/healthcare/working-towards-wellness-business-rationale.jhtml.

31  Business in the Community (BITC) (2011). *The business case for being a responsible business.* UK: BITC.

32  Conference Board: Carroll, A.B. and Shabana K.M. (2011). *The business case for corporate social responsibility.* Director Notes. New York: The Conference Board Inc.

33  IFC (2007). *Stakeholder Engagement: A Good Practice Handbook for Companies Doing Business in Emerging Markets.* Washington DC: IFC.

34  REN21 (2011). *Renewables 2011—Global Status Report.* Paris, Eschborn: REN21 Secretariat and GIZ.

35  See for example, the Principles for Responsible Investment (PRI) initiative [online] where integration of responsible investment principles by over 1000 investment institution signatories stands at approximately US$ 30 trillion. Available at: http://www.unpri.org/about/ [Accessed 20th April 2012].

36  Ibid.

37  GlobeScan, UNEP and SustainAbility (2012). *The GlobeScan/SustainAbility Survey.* Available at: www.sustainability.com. [Accessed 8th May 2012].

38  See The Economics of Ecosystems and Biodiversity (TEEB) (2010). *The Economics of Ecosystems and Biodiversity. Report for Business.* Executive Summary. Available at: http://www.teebweb.org/. See also: UNEP (2010). *Are you a green leader? Business and biodiversity: making the case for a lasting solution.* Available at: http://www.unep.fr/scp/publications/.

# THE NEXT INDUSTRIAL REVOLUTION

By Paul Hawken, Amory B. Lovins, and L. Hunter Lovins

*Emerging possibilities—A new type of industrialism—The loss of living systems—Valuing natural capital—The industrial mind-set—The emerging pattern of scarcity—Four strategies of natural capitalism—Radical resource productivity—Putting the couch potato of industrialism on a diet-An economy of steady service and flow—Restoring the basis of life and commerce*

Imagine for a moment a world where cities have become peaceful and serene because cars and buses are whisper quiet, vehicles exhaust only water vapor, and parks and greenways have replaced unneeded urban freeways. OPEC has ceased to function because the price of oil has fallen to five dollars a barrel, but there are few buyers for it because cheaper and better ways now exist to get the services people once turned to oil to provide. Living standards for all people have dramatically improved, particularly for the poor and those in developing countries. Involuntary unemployment no longer exists, and income taxes have largely been eliminated. Houses, even low-income housing units, can pay part of their mortgage costs by the energy they *produce;* there are few if any active landfills; worldwide forest cover is increasing; dams are being dismantled; atmospheric $CO_2$ levels are decreasing for the first time in two hundred years; and effluent water leaving factories is cleaner than the water coming into them. Industrialized countries have reduced resource use by 80 percent while improving the quality of life. Among these technological changes, there are important social changes. The frayed social nets of Western countries have been repaired. With the explosion of family-wage jobs, welfare demand has fallen. A progressive and active union movement has taken the lead to work with business, environmentalists, and government to create "just transitions" for workers as society phases out coal, nuclear energy, and oil. In communities and towns, churches, corporations, and labor groups promote a new living-wage social contract as the least expensive way to ensure the growth and preservation of valuable social capital. Is this the vision of a utopia? In fact, the changes described here could come about in the decades to come as the result of economic and technological trends already in place.

In the next century, as human population doubles and the resources available per person drop by one-half to three-fourths, a remarkable transformation of industry and commerce can occur. Through this transformation, society will be able to create a vital economy that uses radically less material and energy. This economy can free up resources, reduce taxes on personal income, increase per-capita spending on social ills (while simultaneously reducing those ills), and begin to restore the damaged environment of the earth. These necessary changes done properly can promote economic efficiency, ecological conservation, and social equity.

The industrial revolution that gave rise to modern capitalism greatly expanded the possibilities for the material development of humankind. It continues to do so today, but at a severe price. Since the mid-eighteenth century, more of nature has been destroyed than in all prior history. While industrial systems have reached pinnacles of success, able to muster and accumulate human-made capital on vast levels, *natural capital,* on which civilization depends to create economic prosperity, is rapidly declining, [1] and the rate of loss is increasing proportionate to gains in material well-being. *Natural capital* includes all the familiar resources used by humankind: water, minerals, oil, trees, fish, soil, air, et cetera. But it also encompasses living systems, which include grasslands, savannas, wetlands, estuaries, oceans, coral reefs, riparian corridors, tundras, and rainforests. These are deteriorating worldwide at an unprecedented rate. Within these ecological communities are the fungi, ponds, mammals, humus, amphibians, bacteria, trees, flagellates, insects, songbirds, ferns, starfish, and flowers that make life possible and worth living on this planet.

As more people and businesses place greater strain on living systems, limits to prosperity are coming to be determined by natural capital rather than industrial prowess. This is not to say that the world is running out of commodities in the near future. The prices for most raw materials are at a twenty-eight-year low and are still falling. Supplies are cheap and appear to be abundant, due to a number of reasons: the collapse of the Asian economies, globalization of trade, cheaper transport costs, imbalances in market power that enable commodity traders and middlemen to squeeze producers, and in large measure the success of powerful new extractive technologies, whose correspondingly extensive damage to ecosystems is seldom given a monetary value. After richer ores are exhausted, skilled mining companies can now level and grind up whole mountains of poorer-quality ores to extract the metals desired. But while technology keeps ahead of depletion, providing what appear to be ever-cheaper metals, they only appear cheap, because the stripped rainforest and the mountain of toxic tailings spilling into rivers, the impoverished villages and eroded indigenous cultures—all the consequences they leave in their wake—are not factored into the cost of production.

It is not the supplies of oil or copper that are beginning to limit our development but life itself. Today, our continuing progress is restricted not by the number of fishing boats but by the decreasing numbers of fish; not by the power of pumps but by the depletion of aquifers; not by the number of chainsaws but by the disappearance of primary forests. While living systems are the source of such desired materials as wood, fish, or food, of utmost importance are the *services* that they offer, [2] services that are far more critical to human prosperity than are nonrenewable resources. A forest provides .not only the resource of wood but also the services of water storage and flood management. A healthy environment automatically supplies not only clean air and water, rainfall, ocean productivity, fertile soil, and watershed resilience but also such less-appreciated functions as waste processing (both natural and industrial), buffering against the extremes of weather, and regeneration of the atmosphere.

Humankind has inherited a 3.8-billion-year store of natural capital. At present rates of use and degradation, there will be little left by the end of the next century. This is not only a matter of aesthetics and morality, it is of the utmost practical concern to society and all people. Despite reams of press about the state of the environment and rafts of laws attempting to prevent further loss, the stock of nature capital is plummeting and the vital life-giving services that flow from it are critical to our prosperity.

Natural capitalism recognizes the critical interdependency between the production and use of human-made capital and the maintenance and supply of natural capital. The traditional definition of capital is accumulated wealth in the form of investments, factories, and equipment. Actually, an economy needs four types of capital to function properly:

- human capital, in the form of labor and intelligence, culture, and organization
- financial capital, consisting of cash, investments, and monetary instruments
- manufactured capital, including infrastructure, machines, tools, and factories
- natural capital, made up of resources, living systems, and ecosystem services

The industrial system uses the first three forms of capital to transform natural capital into the stuff of our daily lives: cars, highways, cities, bridges, houses, food, medicine, hospitals, and schools.

The climate debate is a public issue in which the assets at risk are not specific resources, like oil, fish, or timber, but a life-supporting system. One of nature's most critical cycles is the continual exchange of carbon dioxide and oxygen among plants and animals. This "recycling service" is provided by nature free of charge. But today carbon dioxide is building up in the atmosphere, due in part to combustion of fossil fuels. In effect, the capacity of the natural system to recycle carbon dioxide has been exceeded, just as overfishing can exceed the capacity of a fishery to replenish stocks. But what is especially important to realize is that there is no known alternative to nature's carbon cycle service.

Besides climate, the changes in the biosphere are widespread. In the past half century, the world has a lost a fourth of its topsoil and a third . of its forest cover. At present rates of destruction, we will lose 70 percent of the world's coral reefs in our lifetime, host to 25 percent of marine life.[3] In the past three decades, one-third of the planet's resources, its. "natural wealth," has been consumed. We are losing freshwater ecosystems at the rate of 6 percent a year, marine ecosystems by 4 percent a year.[4] There is no longer any serious scientific dispute that the decline in every living system in the world is reaching such levels that an increasing number of them are starting to lose, often at a pace accelerated by the interactions of their decline, their assured ability to sustain the continuity of the life process. We have reached an extraordinary threshold.

Recognition of this shadow side of the success of industrial production has triggered the second of the two great intellectual shifts of the late twentieth century. The end of the Cold War and the fall of communism was the first such shift; the second, now quietly emerging, is the end of the war against life on earth, and the, eventual ascendance of what we call natural capitalism.

Capitalism, as practiced, is a financially profitable, nonsustainable aberration in human development. What might be called "industrial capitalism" does not fully conform to its own accounting principles. It liquidates its capital and calls it income. It neglects to assign any value to the largest stocks of capital it employs—the natural resources and living systems, as well as the social and cultural systems that are the basis of human capital.

But this deficiency in business operations cannot be corrected simply by assigning monetary values to natural capital, for three reasons. First, many of the services we receive from living systems have no known substitutes at any price; for example, oxygen production by green plants. This was demonstrated memorably in 1991–93 when the scientists operating the $200 million Biosphere 2 experiment in Arizona discovered that it was unable to maintain life-supporting oxygen levels for the eight people living inside. Biosphere 1, a.k.a. Planet Earth, performs this task daily at no charge for 6 billion people.

Second, valuing natural capital is a difficult and imprecise exercise at best. Nonetheless, several recent assessments have estimated that biological services flowing directly into society from the stock of natural capital are worth at least $36 trillion annually.[5] That figure is close to the annual gross world product of approximately $39 trillion—a striking measure of the value of natural capital to the economy. If natural capital stocks were given a monetary value, assuming the assets yielded "interest" of $36 trillion annually, the world's natural capital would be valued at somewhere between $400 and $500 trillion—tens of thousands of dollars for every person on the planet. That is

undoubtedly a conservative figure given the fact that anything we can't live without and can't replace at any price could be said to have an infinite value.

Additionally, just as technology cannot replace the planet's life-support systems, so, too, are machines unable to provide a substitute for human intelligence, knowledge, wisdom, organizational abilities, and culture. The World Bank's 1995 *Wealth Index* found the sum value of human capital to be three times greater than all the financial and manufactured capital reflected on global balance sheets.[6] This, too, appears to be a conservative estimate, since it counts only the market value of human employment, not compensated effort or cultural resources.

It is not the aim to assess how to determine value for such unaccounted-for forms of capital. It is clear, however, that behaving as though they are valueless has brought us to the verge of disaster. But if it is in practice difficult to tabulate the value of natural and human capital on balance sheets, how can governments and conscientious businesspersons make decisions about the responsible use of earth's living systems?

## CONVENTIONAL CAPITALISM

Following Einstein's dictum that problems can't be solved within the mind-set that created them, the first step toward any comprehensive economic and ecological change is to understand the mental model that forms the basis of present economic thinking. The mind-set of the present capitalist system might be summarized as follows:

- Economic progress can best occur in free-market systems of production and distribution where reinvested profits make labor and capital increasingly productive.

- Competitive advantage is gained when bigger, more efficient plants manufacture more products for sale to expanding markets.
- Growth in total output (GOP) maximizes human well-being.
- Any resource shortages that do occur will elicit the development of substitutes.
- Concerns for a healthy environment are important but must be balanced against the requirements of economic growth, if a high standard of living is to be maintained.
- Free enterprise and market forces will allocate people and resources to their highest and best uses.

The origins of this worldview go back centuries, but it took the industrial revolution to establish it as the primary economic ideology. This sudden, almost violent, change in the means of production and distribution of goods, in sector after economic sector, introduced a new element that redefined the basic formula for the creation of material products: Machines powered by water, wood, charcoal, coal, oil, and eventually electricity accelerated or accomplished some or all of the work formerly performed by laborers. Human productive capabilities began to grow exponentially. What took two hundred workers in 1770 could be done by a single spinner in the British textile industry by 1812. With such astonishingly improved productivity, the labor force was able to manufacture a vastly larger volume of basic necessities like cloth at greatly reduced cost. This in turn rapidly raised standards of living and real wages, increasing demand for other products in other industries. Further technological breakthroughs proliferated, and as industry after industry became mechanized, leading to even lower prices and higher incomes, all of these factors fueled a self-sustaining and increasing demand for transportation, housing,

education, clothing, and other goods, creating the foundation of modern commerce.[7]

The past two hundred years of massive growth in prosperity and manufactured capital have been accompanied by a prodigious body of economic theory analyzing it, all based on the fallacy that natural and human capital have little value as compared to final output. In the standard industrial model, the creation of value is portrayed as a linear sequence of extraction, production, and distribution: Raw materials are introduced. (Enter nature, stage left.) Labor uses technologies to transform these resources into products, which are sold to create profits. The wastes from production processes, and soon the products themselves, are somehow disposed of somewhere else. (Exit waste, stage right.) The "somewheres" in this scenario are not the concern of classical economics: Enough money can buy enough resources, so the theory goes, and enough "elsewheres" to dispose of them afterward.

This conventional view of value creation is not without its critics. Viewing the economic process as a disembodied, circular flow of value between production and consumption, argues economist Herman Daly, is like trying to understand an animal only in terms of its circulatory system, without taking into account the fact it also has a digestive tract that ties it firmly to its environment at both ends. But there is an even more fundamental critique to be applied here, and it is one based on simple logic. The evidence of our senses is sufficient to tell us that all economic activity—all that human beings are, all that they can ever accomplish—is embedded within the workings of a particular planet. That planet is not growing, so the somewheres and elsewheres are always with us. The increasing removal of resources, their transport and use, and their replacement with waste steadily erodes our stock of natural capital.

With nearly ten thousand new people arriving on earth every hour, a new and unfamiliar pattern of scarcity is now emerging. At the beginning of the industrial revolution, labor was overworked and relatively scarce (the population was about one-tenth of current totals), while global stocks of natural capital were abundant and unexploited. But today the situation has been reversed: After two centuries of rises in labor productivity, the liquidation of natural resources at their extraction cost rather than their replacement value, and the exploitation of living systems as if they were free, infinite, and in perpetual renewal, it is people who have become an abundant resource, while *nature* is becoming disturbingly scarce.

Applying the same economic logic that drove the industrial revolution to this newly emerging pattern of scarcity implies that, if there is to be prosperity in the future, society must make its use of *resources* vastly more productive—deriving four, ten, or even a hundred times as much benefit from each unit of energy, water, materials, or anything else borrowed from the planet and consumed. Achieving this degree of efficiency may not be as difficult as it might seem because from a materials and energy perspective, the economy is massively inefficient. In the United States, the materials used by the metabolism of industry amount to more than twenty times every citizen's weight per day—more than one million pounds per American per year. The global flow of matter, some 500 billion tons per year, most of it wasted, is largely invisible. Yet obtaining, moving, using, and disposing of it is steadily undermining the health of the planet, which is showing ever greater signs of stress, even of biological breakdown. Human beings already use over half the world's accessible surface freshwater have transformed one-third to one-half of its land surface, fix more nitrogen than do all natural systems on land, and appropriate more than two-fifths of the planet's

entire land-based primary biological productivity.[8] The doubling of these burdens with rising population will displace many of the millions of other species, undermining the very web of life.

The resulting ecological strains are also causing or exacerbating many forms of social distress and conflict. For example, grinding poverty, hunger, malnutrition, and rampant disease affect one-third of the world and are growing in absolute numbers; not surprisingly, crime, corruption, lawlessness, and anarchy are also on the rise (the fastest-growing industry in the world is security and private police protection); fleeing refugee populations have increased throughout the nineties to about a hundred million; over a billion people in the world who need to work cannot find jobs, or toil at such menial work that they cannot support themselves or their families;[9] meanwhile, the loss of forests, topsoil, fisheries, and freshwater is, in some cases, exacerbating regional and national conflicts.

What would our economy look like if it fully valued *all* forms of capital, including human and natural capital? What if our economy were organized not around the lifeless abstractions of neoclassical economics and accountancy but around the biological realities of nature? What if Generally· Accepted Accounting Practice booked natural and human capital not as a free amenity in putative inexhaustible supply but as a finite and integrally valuable factor of production? What if, in the absence of a rigorous way to practice such accounting, companies started to act *as if* such principles were in force? This choice is possible and such an economy would offer a stunning new set of opportunities for all of society, amounting to no less than the *next industrial revolution*.

## CAPITALISM AS IF LIVING SYSTEMS MATTERED

Natural capitalism and the possibility of a new industrial system are based on a very different mind-set and set of values than conventional capitalism. Its fundamental assumptions include the following:

- The environment is not a minor factor of production but rather is "an envelope containing, provisioning, and sustaining the entire economy."[10]
- The limiting factor to future economic development is the availability and functionality of *natural capital,* in particular, life-supporting services that have no substitutes and currently have no market value.
- Misconceived or badly designed business systems, population growth, and wasteful patterns of consumption are the primary causes of the loss of natural capital, and all three must be addressed to achieve a sustainable economy.
- Future economic progress can best take place in democratic, market-based systems of production and distribution in which *all* forms of capital are fully valued, including human, manufactured, financial, and natural capital.
- One of the keys to the most beneficial employment of people, money, and the environment is radical increases in resource productivity.
- Human welfare is best served by improving the quality and flow of desired services delivered, rather than by merely increasing the total dollar flow.
- Economic and environmental sustainability depends on redressing global inequities of income and material well-being.

- The best long-term environment for commerce is provided by true democratic systems of governance that are based on the needs of people rather than business.

There are four central strategies of natural capitalism that are a means to enable countries, companies, and communities to operate by behaving as if all forms of capital were valued. Ensuring a perpetual annuity of valuable social and natural processes to serve a growing population is not just a prudent investment but a critical need in the coming decades. Doing so can avert scarcity, perpetuate abundance, and provide a solid basis for social development; it is the basis of responsible stewardship and prosperity for the next century and beyond.

*1. Radical resource productivity.* Radically increased resource productivity is the cornerstone of natural capitalism because using resources more effectively has three significant benefits: It slows resource depletion at one end of the value chain, lowers pollution at the other end, and provides a basis to increase worldwide employment with meaningful jobs. The result can be lower costs for business and society, which no longer has to pay for the chief causes of ecosystem and social disruption. Nearly all environmental and social harm is an artifact of the uneconomically wasteful use of human and natural resources, but radical resource productivity strategies can nearly halt the degradation of the biosphere, make it more profitable to employ people, and thus safeguard against the loss of vital living systems and social cohesion.

*2. Biomimicry.* Reducing the wasteful throughput of materials—indeed, eliminating the very idea of waste—can be accomplished by redesigning industrial systems on biological lines that change the nature of industrial processes and materials, enabling the constant reuse of materials in continuous closed cycles, and often the elimination of toxicity.

*3. Service and flow economy.* This calls for a fundamental change in the relationship between producer and consumer, a shift from an economy of goods and purchases to one of *service* and *flow*. In essence, an economy that is based on a flow of economic services can better protect the ecosystem services upon which it depends. This will entail a new perception of value, a shift from the acquisition of goods as a measure of affluence to an economy where the continuous receipt of quality, utility, and performance promotes well-being. This concept offers incentives to put into practice the first two innovations of natural capitalism by restructuring the economy to focus on relationships that better meet customers' changing value needs and to reward automatically both resource productivity and closed-loop cycles of materials use.

*4. Investing in natural capital.* This works toward reversing worldwide planetary destruction through reinvestments in sustaining, restoring, and expanding stocks of natural capital, so that the biosphere can produce more abundant ecosystem services and natural resources.

All four changes are interrelated and interdependent; all four generate numerous benefits and opportunities in markets, finance, materials, distribution, and employment. Together, they can reduce environmental harm, create economic growth, and increase meaningful employment.

## RESOURCE PRODUCTIVITY

Imagine giving a speech to Parliament in 1750 predicting that within seventy years human productivity would rise to the point that one person could do the work of two hundred. The speaker would have been branded as daft or worse. Imagine a similar scene today. Experts are testifying in Congress, predicting that we will increase the productivity of our resources in the next seventy years by a factor of four, ten, even one hundred. Just as it was impossible 250 years ago to conceive of an individual's doing two hundred times more work, it is equally difficult for us today to imagine a kilowatt-hour or board foot being ten or a hundred times more productive than it is now.

Although the movement toward radical resource productivity has been under way for decades, its clarion call came in the fall of 1994, when a group of sixteen scientists, economists, government officials, and businesspeople convened and, sponsored by Friedrich Schmidt-Bleek of the Wuppertal Institute for Climate, Environment, and Energy in Germany, published the "Carnoules Declaration." Participants had come from Europe, the United States, Japan, England, Canada, and India to the French village of Carnoules to discuss their belief that human activities were at risk from the ecological and social impact of materials and energy use. The Factor Ten Club, as the group came to call itself, called for a leap in resource productivity to reverse the growing damage. The declaration began with these prophetic words: "Within one generation, nations can achieve a ten-fold increase in the efficiency with which they use energy, natural resources and other materials."[11]

In the years since, Factor Ten (a 90 percent reduction in energy and materials intensity) and Factor Four (a 75 percent reduction) have entered the vocabulary of government officials, planners, academics, and businesspeople throughout the world.[12] The governments of Austria, the Netherlands, and Norway have publicly committed to pursuing Factor Four efficiencies. The same approach has been endorsed by the European Union as the new paradigm for sustainable development. Austria, Sweden, and OECD environment ministers have urged the adoption of Factor Ten goals, as have the World Business Council for Sustainable Development and the United Nations Environment Program

(UNEP).[13] The concept is not only common parlance for most environmental ministers in the world, but such leading corporations as Dow Europe and Mitsubishi Electric see it as a powerful strategy to gain a competitive advantage. Among all major industrial nations, the United States probably has the least familiarity with and understanding of these ideas.

At its simplest, increasing resource productivity means obtaining the same amount of utility or work from a product or process while using less material and energy. In manufacturing, transportation, forestry, construction, energy, and other industrial sectors, mounting empirical evidence suggests that radical improvements in resource productivity are both practical and cost-effective, even in the most modern industries. Companies and designers are developing ways to make natural resources—energy, metals, water, and forests—work five, ten, even one hundred times harder than they do today. These efficiencies transcend the marginal gains in performance that industry constantly seeks as part of its evolution. Instead, *revolutionary* leaps in design and technology will alter industry itself as demonstrated in the following chapters. Investments in the productivity revolution are not only repaid over time by the saved resources but in many cases can *reduce* initial capital investments.

When engineers speak of "efficiency," they refer to the amount of output a process provides per unit of input. Higher efficiency thus means doing more with less, measuring both factors in physical terms. When economists refer to efficiency, however, their definition differs in two ways. First, they usually measure a process or outcome in terms of expenditure of money—how the market value of what was produced compares to the market cost of the labor and other inputs used to create it. Second, "economic efficiency" typically refers to how fully and perfectly market mechanisms

are being harnessed to minimize the monetary total factor cost of production. Of course it's important to harness economically efficient market mechanisms, and we share economists' devotion to that goal. But to avoid confusion, when we suggest using market tools to achieve "resource productivity" and "resource efficiency," we use those terms in the engineering sense.

Resource productivity doesn't just save resources and money; it can also improve the quality of life. Listen to the din of daily existence—the City and freeway traffic, the airplanes, the garbage trucks outside urban windows—and consider this: The waste and the noise are signs of inefficiency, and they represent money being thrown away. They will disappear as surely as did manure from the nineteenth-century streets of London and New York. Inevitably, industry will redesign everything it make and does, in order to participate in the coming productivity revolution. We will be able to see better with resource-efficient lighting systems, produce higher-quality goods in efficient factories, travel more safely and comfortably in efficient vehicles, feel more comfortable (and do substantially more and better work)[14] in efficient buildings, and be better nourished by efficiently grown food. An air-conditioning system that uses 90 percent less energy or a building so efficient that it needs no air-conditioning at all may not fascinate the average citizen, but the fact that they are quiet and produce greater comfort while reducing energy costs should appeal even to technophobes. That such options save money should interest everyone.

The unexpectedly large improvements to be gamed by resource productivity offer an entirely new terrain for business invention, growth, and development. Its advantages can also dispel the long-held belief that core business values and environmental responsibility are incompatible or at odds. In fact, the massive inefficiencies that are causing environmental degradation almost always cost more than the measures that would reverse them.

But even as Factor Ten goals are driving reductions in materials and energy flows, some governments are continuing to create and administer laws, policies, taxes, and subsidies that have quite the opposite effect. Hundreds of billions of dollars of taxpayers' money are annually diverted to promote inefficient and unproductive material and energy use. These include subsidies to mining, oil, coal, fishing, and forest industries as well as agricultural practices that degrade soil fertility and use wasteful amounts of water and chemicals. Many of these subsidies are vestigial, some dating as far back as the eighteenth century, when European powers provided entrepreneurs with incentives to find and exploit colonial resources. Taxes extracted from labor subsidize patterns of resource use that in turn displace workers, an ironic situation that is becoming increasingly apparent and unacceptable, particularly in Europe, where there is chronically high unemployment. Already, tax reforms aimed at increasing employment by shifting taxes away from people to the use of resources have started to be instituted in the Netherlands, Germany, Britain, Sweden, and Denmark, and are being seriously proposed across Europe.

In less developed countries, people need realistic and achievable means to better their lives. The world's growing population cannot attain a Western standard of living by following traditional industrial paths to development, for the resources required are too vast, too expensive, and too damaging to local and global systems. Instead, radical improvements in resource productivity expand their possibilities for growth, and can help to ameliorate the polarization of wealth between rich and poor segments of the globe. When the world's nations met in Brazil at the Earth Summit in 1992 to discuss the environment and human development, some treaties and proposals proved to be highly divisive because it appeared that they put a lid on the ability of nonindustrialized countries to pursue development. Natural capitalism provides a practical agenda for development wherein the actions of both developed and developing nations are mutually supportive.

## BIOMIMICRY

To appreciate the potential of radical resource productivity, it is helpful to recognize that the present industrial system is, practically speaking, a couch potato: It eats too much junk food and gets insufficient exercise. In its late maturity, industrial society runs on life-support systems that require enormous heat and pressure, are petrochemically dependent and materials-intensive, and require large flows of toxic and hazardous chemicals. These industrial "empty calories" end up as pollution, acid rain, and greenhouse gases, harming environmental, social, and financial systems. Even though all the reengineering and downsizing trends of the past decade were supposed to sweep away corporate inefficiency, the U.S. economy remains astoundingly inefficient: It has been estimated that only 6 percent of its vast flows of materials actually end up in products.[15] Overall, the ratio of waste to the *durable* products that constitute material wealth may be closer to one hundred to one. The whole economy is less than 10 percent—probably only a few percent—as energy-efficient as the laws of physics permit.[16]

This waste is currently rewarded by deliberate distortions in the marketplace, in the form of policies like subsidies to industries that extract raw materials from the earth and damage the biosphere. As long as that damage goes unaccounted for, as long as virgin resource prices are maintained at artificially low levels, it makes sense to continue to use virgin materials rather than reuse resources discarded from previous products. As long as it is assumed that there are "free goods" in the world—pure water, clean air, hydrocarbon combustion, virgin forests, veins of minerals—large-scale, energy-and materials-intensive manufacturing methods will dominate, and labor will be increasingly marginalized.[17] In contrast, if the subsidies distorting resource prices were removed or reversed, it would be advantageous to employ more people and use fewer virgin materials.

Even without the removal of subsidies, the economics of resource productivity are already encouraging industry to reinvent itself to be more in accord with biological systems. Growing competitive pressures to save resources are opening up exciting frontiers for chemists, physicists, process engineers, biologists, and industrial designers. They are reexamining the energy, materials, and manufacturing systems required to provide the specific qualities (strength, warmth, structure, protection, function, speed, tension, motion, skin) required by products and end users and are turning away from mechanical systems requiring heavy metals, combustion, and petroleum to seek solutions that use minimal inputs, lower temperatures, and enzymatic reactions. Business is switching to imitating biological and ecosystem processes replicating natural methods of production and engineering to manufacture chemicals, materials, and compounds, and soon maybe even microprocessors. Some of the most exciting developments have resulted from emulating nature's life-temperature, low-pressure, solar-powered assembly techniques, whose products rival anything human-made. Science writer Janine Benyus points out that spiders make silk, strong as Kevlar but much tougher, from digested crickets and flies, without needing boiling sulfuric acid and high-temperature extruders. The abalone generates an inner shell twice as tough

as our best ceramics, and diatoms make glass, both processes employing seawater with no furnaces. Trees turn sunlight, water, and air into cellulose, a sugar stiffer and stronger than nylon, and bind it into wood, a natural composite with a higher bending strength and stiffness than concrete or steel. We may never grow as skillful as spiders, abalone, diatoms, or trees, but smart designers are apprenticing themselves to nature to learn the benign chemistry of its processes.

Pharmaceutical companies are becoming microbial ranchers managing herds of enzymes. Biological farming manages soil ecosystems in order to increase the amount of biota and life per acre by keen knowledge of food chains, species interactions, and nutrient flows, minimizing crop losses and maximizing yields by fostering diversity. Meta-industrial engineers are creating "zero-emission," industrial parks whose tenants will constitute an industrial ecosystem in which one company will feed upon the nontoxic and useful wastes of another. Architects and builders are creating structures that process their own wastewater, capture light, create energy, and provide habitat for wildlife and wealth for the community, all the while improving worker productivity, morale, and health.[18] High-temperature, centralized power plants are starting to be replaced by smaller-scale, renewable power generation. In chemistry, we can look forward to the end of the witches' brew of dangerous substances invented this century, from DDT, PCB, CFCs, and Thalidomide to Dieldrin and xeno-estrogens. The eighty thousand different chemicals now manufactured end up everywhere, as Donella Meadows remarks, from our "stratosphere to our sperm." They were created to accomplish functions that can now be carried out far more efficiently with biodegradable and naturally occurring compounds.

## SERVICE AND FLOW

Beginning in the mid-1980s, Swiss industry analyst. Walter Stahel and German chemist Michael Braungart independently proposed a new industrial model that is now gradually taking shape. Rather than an economy in which *goods* are made and sold, these visionaries imagined a *service economy* wherein consumers obtain *services* by leasing or renting goods rather than buying them outright. (Their plan should not be confused with the conventional definition of a service economy, in which burger-flippers outnumber steelworkers.) Manufacturers cease thinking of themselves as sellers of products and become, instead, deliverers of service, provided by long-lasting, upgradeable durables. Their goal is selling results rather than equipment, performance and satisfaction rather than motors, fans, plastics, or condensers.

The system can be demonstrated by a familiar example. Instead of purchasing a washing machine, consumers could pay a monthly fee to obtain the *service* of having their clothes cleaned. The washer would have a counter on it, just like an office photocopier, and would be maintained by the manufacturer on a regular basis, much the way mainframe computers are. If the machine ceased to provide its specific service, the manufacturer would be responsible for replacing or repairing it at no charge to the customer, because the washing machine would remain the property of the manufacturer. The concept could likewise be applied to computers, cars, VCRs, refrigerators, and almost every other durable that people now buy, use up, and ultimately throw away. Because products would be returned to the manufacturer for continuous repair, reuse, and remanufacturing, Stahel called the process "cradle-to-cradle" [19]

Many companies are adopting Stahel's principles. Agfa Gaevert pioneered the leasing of copier services, which spread to the entire industry.[20] The Carrier Corporation, a division of United Technologies, is creating a program to sell coolth (the opposite of warmth) to companies while retaining ownership of the air-conditioning equipment. The Interface Corporation is beginning to lease the warmth, beauty, and comfort of its floor-covering services rather than selling carpets.

Braungart's model of a *service economy* focuses on the nature of material cycles. In this perspective, if a given product lasts a long time but its waste materials cannot be reincorporated into new manufacturing or biological cycles, then the producer must accept responsibility for the waste with all its attendant problems of toxicity, resource overuse, worker safety, and environmental damage. Braungart views the world as a series of metabolisms in which the creations of human beings, like the creations of nature, become "food" for interdependent systems, returning to either an industrial or a biological cycle after their useful life is completed. To some, especially frugal Scots and New Englanders, this might not sound a novel concept at all. Ralph Waldo Emerson once wrote, "Nothing in nature is exhausted in its first use. When a thing has served an end to the uttermost, it is wholly new for an ulterior service."[21] In simpler times, such proverbial wisdom had highly practical applications. Today, the complexity of modern materials makes this almost impossible. Thus, Braungart proposed an Intelligent Product System whereby those products that do not degrade back into natural nutrient cycles be designed so that they can be deconstructed and completely reincorporated into *technical nutrient* cycles of industry.[22]

Another way to conceive of this method is to imagine an industrial system that has no provision for landfills, outfalls, or smokestacks. If a company knew that nothing that came into its factory could be thrown away, and that everything it produced would eventually return, how would it design its components and products? The question is more than a theoretical construct, because the earth works under precisely these strictures.

In a *service economy,* the product is a means, not an end. The manufacturer's leasing and ultimate recovery of the product means that the product remains an asset. The minimization of materials use, the maximization of product durability, and enhanced ease of maintenance not only improve the customer's experience and value but also protect the manufacturer's investment and hence its bottom line. *Both* producer and customer have an incentive for continuously improving resource productivity, which in turn further protects ecosystems. Under this shared incentive, both parties form a relationship that continuously anticipates and meets the customer's evolving value needs— and meanwhile rewards both parties for reducing the burdens on the planet.

The service paradigm has other benefits as well: It increases employment, because when products are designed to be reincorporated into manufacturing cycles, waste declines, and demand for labor increases. In manufacturing, about one-fourth of the labor force is engaged in the fabrication of basic raw materials such as steel, glass, cement, silicon, and resins, while three-quarters are in the production phase. The reverse is true for energy inputs: Three times as much energy is used to extract virgin or primary materials as is used to manufacture products from those materials. Substituting reused or more durable manufactured goods for primary materials therefore uses less energy but provides more jobs. [23]

An economy based on a service-and-flow model could also help stabilize the business cycle, because customers would be purchasing flows of services, which they need continuously, rather than durable equipment that's affordable only in good years. Service providers would

have an incentive to keep their assets productive for as long as possible, rather than prematurely scrapping them in order to sell replacements. Over-and undercapacity would largely disappear, as business would no longer have to be concerned about delivery or backlogs if it is contracting from a service provider. Gone would be end-of-year rebates to move excess automobile inventory, built for customers who never ordered them because managerial production quotas were increased in order to amortize expensive capital equipment that was never needed in the first place. As it stands now, durables manufacturers have a love-hate relationship with durability. But when they become service providers, their long-and short-term incentives become perfectly attuned to what customers want, the environment deserves, labor needs, and the economy can support. [24]

## INVESTING IN NATURAL CAPITAL

When a manufacturer realizes that a supplier of key components is overextended and running behind on deliveries, it takes immediate action lest its own production lines come to a halt. Living systems are a supplier of key components for the life of the planet, and they are now falling behind on their orders. Until recently, business could ignore such shortages because they didn't affect production and didn't increase costs. That situation may be changing, however, as rising weather-related claims come to burden insurance companies and world agriculture. (In 1998, violent weather caused upward of $90 billion worth of damage worldwide, a figure that represented more weather-related losses than were accounted for through the entire decade of the 1980s. The losses were greatly compounded by deforestation and climate change, factors that increase the frequency and severity of disasters. In human terms, 300 million people were permanently or temporarily displaced from their homes; this figure includes the dislocations caused by Hurricane Mitch, the deadliest Atlantic storm in two centuries.)[25] If the flow of services from industrial systems is to be sustained or increased in the future for a growing population, the vital flow of life-supporting services from living systems will have to be maintained and increased. For this to be possible will require investments in natural capital.

As both globalization and Balkanization proceed, and as the per-capita availability of water, arable land, and fish continue to decline (as they have done since 1980), the world faces the danger of being torn apart by regional conflicts instigated at least in part by resource shortages or imbalances and associated income polarization.[26] Whether it involves oil[27] or water[28], cobalt or fish, access to resources is playing an ever more prominent role in generating conflict. In addition, many social instabilities and refugee populations—twelve million refugees now wander the world—are created or worsened by ecological destruction, from Haiti to Somalia to Jordan. On April 9, 1996, Secretary of State Warren Christopher gave perhaps the first speech by an American cabinet officer that linked global security with the environment. His words may become prophetic for future foreign policy decisions: "... [E]nvironmental forces transcend borders and oceans to threaten directly the health, prosperity and jobs of American citizens. ... [A]ddressing natural resource issues is frequently critical to achieving political and economic stability, and to pursuing our strategic goals around the world"

Societies need to adopt shared goals that enhance social welfare but that are not the prerogatives of specific value or belief systems. Natural capitalism is one such objective. It is neither conservative nor liberal in its ideology, but appeals to both constituencies. Since it is a means, and not an end, it doesn't advocate a particular social outcome but rather makes possible many different ends. Therefore, whatever the various visions different parties or factions espouse, society can work toward resource productivity now, without waiting to resolve disputes about policy.

Engineers have already designed hydrogen-fuel-cell-powered cars to be plug-in electric generators that may become the power plants of the future. Buildings already exist that make oxygen, solar power, and drinking water and can help pay the mortgage while their tenants work inside them. Deprintable and reprintable papers and inks, together with other innovative ways to use fiber, could enable the world's supply of lumber and pulp to be grown in an area about the size of Iowa. Weeds can yield potent pharmaceuticals; cellulose-based plastics have been shown to be strong, reusable, and compostable; and luxurious carpets can be made from land-filled scrap. Roofs and windows, even roads, can do double duty as solar-electric collectors, and efficient car-free cities are being designed so that men and women no longer spend their days driving to obtain the goods and services of daily life. These are among the thou-sands of innovations that are resulting from natural capitalism.

Many of the techniques and methods described here can be used by individuals and small businesses. Other approaches are more suitable for corporations, even whole industrial sectors; still others better suit local or central governments. Collectively, these techniques offer a powerful menu of new ways to make resource productivity the foundation of a lasting and prosperous economy—from Main Street to Wall Street, from your house to the White House, and from the village to the globe.

Although there is an overwhelming emphasis on what we do with our machines, manufacturing processes, and materials, its purpose is to support the human community and all life-support systems. There is a large body of literature that addresses the nature of specific living systems, from coral reefs to estuarine systems to worldwide topsoil formation. Our focus is to bring about those changes in the human Side of the economy that can help preserve and reconstitute these systems, to try and show for now and all time to come that there is no true separation between how we support life economically and ecologically.

Section VIII

# PERSONAL BEHAVIOR CHOICES

# MAKING ENVIRONMENTAL HISTORY

By James Farrell

*It seems pathetic that it has to be us, with all the other citizens of the planet, and all the other resources out there, but since no one else is doing anything about it, we don't really have a choice.*
Jerry Garcia, "Interview: Jerry Garcia (1989):' *High Times*

*A new idea is first condemned as ridiculous, then dismissed as trivial, until finally it becomes what everybody knows.*
William James

*We cannot expect to be able to solve any complex problems from within the same state of consciousness that created them.*
Albert Einstein

*Never doubt for a moment that a small group of dedicated citizens can change*
*the world. Indeed, it's the only-thing that ever has.*
Margaret Mead

The nature of college is a part of the nature of American life, which is just a small part of nature. Students live on specific college campuses, but they also live in American environmental history. The ideas, institutions, languages, and assumptions of our world, including campus culture, come to us from the minds and the practices of the people of the past. Institutions like capitalism and the corporation, advertising and marketing, religion and schools, malls and supermarkets, industry and industrial agriculture all come from the people who preceded us. These traditions are the default settings of American culture and its college culture.

On campus every day, students make environmental history by deferring to these default settings or deviating from them. Repeating the routines of their days and nights, students unconsciously make history by voting for more of the same, living the common sense of their culture. Until recently, common sense seemed to make sense because it contributed to "progress" and "success." Now, however, some of that progress seems questionable, and some of our cultural successes are leading to ecological failures. The assumptions and expectations of the twentieth century won't work in the twenty-first century.

It's time, therefore, to make a different kind of environmental history. Conscious of the complexity of our lives and our complicity with systems that we really don't support, we have a chance to change our ideas and institutions, our habits and our habitat. We have a chance to change the world into a place we will love to live in, a place we can proudly pass on to our children and our children's children.

## MAKING HISTORY ON CAMPUS: WHAT'S HAPPENING

Thousands of American colleges have already started to change the world by changing the standard operating procedures of their campuses. Pledging eventual carbon neutrality, for example, signers of the American College and University Presidents' Climate Commitment (ACUPCC)—have noted:

> America's higher education community can play a determinant role in addressing climate change. Leading society in this effort fits squarely into the educational, research, and public service missions of higher education. No other institution in society has the influence, the critical mass, and the diversity of skills needed to successfully reverse global warming.

Tomorrow's architects, engineers, attorneys, business leaders, scientists, urban planners, policy analysts, cultural leaders, journalists, advocates, activists, and politicians—more than seventeen million of them—are currently attending the more than four thousand institutions of higher learning in the United States. Higher education is also a $317 billion economic engine that employs millions of people and spends billions of dollars on fuel, energy, products, services and infrastructure.[1]

Such commitments have made college campuses into a powerful proving ground for the ideas and innovations of the ecological revolution. From presidential suites to custodial closets, college personnel are proving that it's possible to institutionalize more sustainable solutions to perennial problems. Food services like Bon Appétit, for example, have committed to serving fresh, responsibly-raised food on its college campuses. Facilities directors all over the country are transforming campus infrastructure to save energy and money by establishing wind turbines, solar arrays, dean-energy purchases, and carbon offsets. They're commissioning green buildings from innovative architects and are looking for ways to heat and cool buildings more efficiently. For commuters, colleges are offering free transit passes, bike-sharing programs, parking preferences for hybrid or fuel-efficient vehicles (or both), and campus fleets that run on biodiesel or other alternative fuels. Many colleges are hiring sustainability coordinators to keep an eye on all the different possibilities for doing good by doing better, and to share successes both with the college community and other interested parties.

But presidents and administrators are not the only impetus for these positive transformations.

Students and student organizations foment enormous change on almost every college—campus, initiating and supporting a wide variety of projects and policies. Voting to increase student fees to increase sustainability efforts, creating CERFs—Clean Energy Revolving Funds—to capitalize on energy-saving projects on campuses, starting their own organic farms to supply the college food service, and banding together in college coalitions to sponsor recycling competitions or "energy wars," these students are revolutionizing college life and having a lot of fun doing it.[2] Additionally, organizations like the public interest research groups (PIRGs) and the Energy Action Coalition are equipping students with skills to enhance their political clout on and off campus.[3]

And yet, even this unprecedented positive activity it is not enough. For the most part, there's still more consciousness raising than cultural change. Even as environmental policies and participation change, there's still not much change in everyday life. The *culture* of college isn't changing much at all, and neither is American culture.

## MAKING HISTORY ON CAMPUS: WHAT'S NOT HAPPENING

Despite all of this work, college students still live like most Americans, and they graduate into a world that's simply not sustainable. They live in climate-controlled dorms that still affect the climate, and they fill their spaces with stuff they "need," discarding it later as if it wasn't worth much—which, of course, it's not, in monetary terms. Like other Americans, they're nourished by nature and agriculture, but they don't return the favor. Driving their cars to a variety of destinations, they're also driving to a new world of reduced resources and enhanced environmental dangers. Students continue to pursue the pleasures of private life, opting for a world in which those pleasures will be increasingly rare. Watching TV and YouTube, checking Facebook and playing video games, they're screening themselves from the serious lessons the world might offer. Conforming to consumer culture, they're consuming the future. Campus life like this is dangerous because it teaches students to think that American consumer culture is normal, natural, normative, and inevitable. And if today's college students graduate with that understanding, the "real world" of our future will be hostile to many of their hopes and plans.

It's not entirely by accident, of course, that college culture is primarily consumer culture. Advertisers, marketers, and retailers work very hard to shape students' environmental values (and thus environmental impacts) by shaping their habits of consumption. They know that college is one of the best places to establish those habits for the rest of their lives. And as long as they continue to buy into conceptions of the good life that marketers supply, they'll keep filling students' minds with desire, their rooms with stuff, and their planet with problems.[4]

Ironically the problem isn't failure; it's a certain kind of success. The problem isn't evil people doing devilish things. It's normal people—us—living ordinary lives. We've succeeded in creating a consumer society that maximizes the pursuit of happiness—especially the material pursuits. But this success is also a failure, because, as Bill McKibben has suggested, "our environmental problems come from normal human life, but there are so many of us living those normal lives that something abnormal is happening.[5]

We're caught in what David Orr calls a "social trap," a situation in which people are drawn into individually rational behavior that is destructive to the planet. The environmental crisis is like a traffic jam. No single driver caused it, but every driver contributes. No one person can eliminate the problem alone, but a system of restraint and regulation can improve it. When individual actions—like driving—increase global warming, solutions must be both individual *and* social, both personal *and* political. Sustainability isn't a technical problem, it's a cultural problem, and we are that culture.[6]

Sadly, college culture is lagging behind the architectural and technological accomplishments on college campuses. In daily routines, students today still live pretty much like their predecessors a generation or two ago. So the challenge for students is to develop a college culture that's appropriate to the environmental realities of our time. There is a lot of serious thinking to do, and not just the standard thinking regarding schoolwork, jobs, family, and friends. In a world of myriad possibilities, students finally get to tackle some bigger questions, whether or not professors ever require it, including the difference between a good job and good work, and the relationship of fun and fulfillment. On campus and off, students can contemplate how college culture and consumption affects nature, and what they plan to do about it. Setting aside the technological determinism of American culture, they can think twice about tools and technologies, wondering whether cars, computers, and contrivances can ever get them where they really want to go. In their personal lives, they need to think about what they mean by love, *how* they mean to love, and whether their intimate love has any connections to larger human and natural communities. Sooner or later, too, they'll confront questions of ultimate—meaning, their own spirituality, and the coping and hoping mechanisms that will help them brave the new world. Living away from home, students will have new resources to think about self and society and how individuals participate in communities, human and natural. In short, students get to think twice about America's commonsense environmental values and some *uncommonsense* alternatives. This thinking will be work, but it will be good work, and as students convert ideals to action in their lives, it will free them to live the lives they want to lead in a society that supports their deepest values instead of undermining them.

## MAKING ENVIRONMENTAL HISTORY: WHAT COULD HAPPEN

The culture we live in isn't appropriate for the twenty-first century. Unfortunately, cultures aren't like the new things we buy at the mall Instead, like organisms, cultures evolve, with mutations here and there that are either accepted as a new common sense or discarded as simply deviant. Some of the mutations are almost accidental, as people adopt new technologies and adapt to them. Philo Farnsworth didn't intend to change American culture when he invented the first functional television system, but he did. Other mutations are very intentional. When Americans of the past argued for a new nation, an end to slavery, equality for women, or an end to smallpox or malaria, they fought hard for a better culture.

On our campuses, students have experienced changes in college culture that have arrived with technological innovations like computers, cell phones, and iPods, but they haven't seen many cultural changes that might support a sustainable society. So, as Gandhi is supposed to have said, "We must be the change we wish to see in the world." But students must also promote the change they hope to see on campus so that others will change as well. To that end, we need to think carefully about the roots and routes of cultural change, asking what makes people change their minds and their lives.

This won't be easy. Summarizing his research in *The First Year out,* Tim Clydesdale reminds us:

We are, in the end, deeply cultural beings. Mainstream. American teens have largely become what popular American moral culture has shaped them to become. They form more or less successful patterns of navigating relationships and managing gratifications, the prefer consumptive leisure and willingly insert themselves into the work-and-spend cycle, they pursue the practical educational credentials necessary to sustain these

## American (Environmental) Values

| COMMON SENSE | UNCOMMON SENSE |
| --- | --- |
| Individualism | "Land Community" |
| Anthropocentrism | Biocentrism |
| Just us | Justice |
| Freedom—independence | Responsibility—interdependence |
| Convenience—time poverty | Conviviality—time affluence, timeliness |
| Indoorness | Outdoorness |
| Resourcism | Ecosystem services |
| Domination | Right relation |
| Dominion | Stewardship |
| "Back to nature" | Always in nature |
| *Homo economicus* | *Homo ecologicus* |
| Life of abundance | Abundant life |
| Materialism—desire for stuff | Dematerialization—desire for fulfillment |
| Buyosphere | Biosphere |
| Quantity of life | Quality of life |
| Cheapness | Value |
| More forme | Enough for everybody |
| Neophilia—love of novelty | Biophilia—love of life |
| Democracy of goods | Democracy of common good |
| Me, the people | We, the people |
| Culture of distraction | Culture of concentration |
| Fun-damentalism | Fulfillment |
| Globalism | Localism |
| Human logic | Bio-logic |
| Hubris | Humility |
| Efficiency—doing things right | Efficacy—doing the right things |
| Presentism | Posterity |
| Nationalism—centralization | Localism—decentralization |
| Globalization—global capitalism | Planetization—natural capitalism |
| Out of sight, out of mind | On-site, in mind |
| Fossil fuelishness | Renewable energies |
| "The good life" | A better life |
| Progress | Progress |

patterns and preferences, and they give little attention to that which lies beyond their microworlds. Most teens, moreover, ignore. (if not resist) all opportunities to critically evaluate these patterns or to understand the wider world. To do so, they use an identity lockbox.[7]

Opening that lockbox—the lockbox of the mind—is a challenge. It means being counter-cultural in the most profound sense. Fortunately for college students, college is a perfect place for learning a new rhetoric of hope and trans-formation, and learning how to transform that rhetoric into reality. Rhetoric is the art of helping people hear the truths they need to know, and living those truths day after day. And hope isn't just wishful thinking. Hope ends in action. Hope is not just something to have, it's something to do.[8]

## COMMONS SENSE FOR COLLEGE CULTURE

There are many different ways of changing college culture and countless ways to do things right. There are books that offer advice on sustainable behavior. Some useful basic guides include *Wake Up and Smell the Planet,* by *Grist,* and *Green Living,* by *E!* magazine. Older (and wiser) is *The Consumer's Guide to Effective Environmental Choices,* by Michael Brown, which makes it clear that political choices *are* con-sumer choices because they set the parameters for production, and require the information that consumers need to make effective envi-ronmental choices. On the Web, there's useful and provocative information on almost any environmental topic: The Sierra Club maintains a "Green Tips Library," and *National Geographic* offers a green guide. The Worldwatch Institute offers *Good Stuff? A Behind-the-Scenes Guide to*

*the Things We Buy,* and the Center for a New American Dream challenges us with a variety of green initiatives. There are also a number of green-living guides at particular colleges and universities.

---

**Commons sense**: The common sense of the twenty-first century, in which "everybody knows" that human life depends on other lives in the biosphere—and the health of the biosphere it self.

---

In looking forward to a new culture of col-lege, there's no need to repeat here the good advice that's already available in abundance. But it might be worthwhile to consider a few general principles of a new common sense that would connect our lives to our deepest values. The new common sense of college culture will be a *commons* sense, a set of beliefs and customs that accustom us to our creative and conserving role in the global commons. In colonial New England, the commons was the area in which residents could commonly pas-ture their animals, the space that wasn't private property but was shared usefully by the people of the town. In the twenty-first century, our commons is the biosphere and its biodiversity, the source of the resources—earth, water, and air, interacting with the fire of the sun—that sustain us. In the twentieth century, we treated the commons like it was infinite—and infinitely resilient. In the twenty-first century, we'll need to treat it uncommonly well because unless we conserve and care for it, the commons will be exhausted. On college campuses, therefore, the new "commons sense" will help us live our lives so that the tragedy of the commons becomes the triumph of the commons. Here, then, are some guidelines for a "commons sense" of col-lege culture:

1.  College education isn't just classes, papers, and GPAs. It's also an open invitation to engage designing minds, first in understanding the designs of nature, second in understanding the culture of nature, and finally in designing a culture that enriches nature's health and our own deep-fulfillment. William McDonough contends, "Design is the first signal of human intention. Our goal is a delightfully diverse, safe, healthy, and just world, with clean air, water, soil, and power—economically, equitably, ecologically, and elegantly enjoyed." College is where we learn how to live well by design. It's where students can learn what in the world they're good at, and how they might be good for the world.[9]

2.  John Dewey said, "Education isn't preparation for life. Education is life itself." At college, students practice academic disciplines in their classes, but they're also practicing human disciplines in everyday life. Because it's a culture committed at least theoretically to mindfulness; college is the right time to establish regenerative routines for the real world, developing habits that enrich habitats. It's a good place to make the mistakes that inevitably come with innovation, and to learn from them.[10]

3.  Humans are solar-powered people on a solar-powered planet. College culture needs to show American culture how to adjust to that reality. On campus, students can begin to opt out of cycles that depend on the ancient sunshine of fossil fuels and opt in to designs that power life with current sunshine—solar power, wind power, and biomass. In their daily living they can develop models for a culture of permanence by purchasing megawatts of clean energy and generating *negawatts* of energy through conservation—almost always the most efficient energy strategy. The cleanest energy is still the energy we don't need, so we need to think carefully about the nature of needs and the nature of enough. As Bill McKibben says, "How much is enough?" is the most important question of our time—even if it's not on the final exam.[11]

4.  Students are embodied beings in a material world, and therefore their materialism matters. Right now, most Americans practice a materialism that shows little regard for the materials we use in our lives or for the deep satisfactions that better designs might offer. We need a new materialism based on a reverence for the physical world and committed to using materials only for our essential human needs. And college, where all we really need is books, some food, a bed, and curiosity, is the perfect place to practice. This new materialism will let us dematerialize some of the satisfactions of our lives, and find more fulfillment with less stuff and a smaller ecological footprint. Etymologically, "thrift" and "thriving" are related, and they could be related again in a "commons sense" where the practice of restraint yields an increase in real satisfaction. Giving up, after all, is a form of giving. It's a gift to the future and others in the world because our restraint is somebody else's reprieve. Our sufficiency provides for the future's sufficiency, too.

5.  For many students, college is their first chance to budget money on their own. As such, it is a wonderful opportunity to practice putting their money where their values are. It's a chance to show friends, family, and others how to buy environmentally responsible products and how to buy into systems of sourcing that enhance the environment and the lives of the people we depend on. Both college students and their institutions can leverage purchases to change the nature of the supply chain. As

citizens, students can also work to put their taxes where their values are, creating a government that supports environmental innovation. At college, students have the luxury to advocate for programs and policies that might make taxes more worthwhile, reversing current subsidies for global weirding and biodiversity loss, and advancing the energies of communities that are good for their places and the whole planet.

6. The economy is for people, and not vice versa. Both "economy" and "ecology" come from the Greek root "oikos," our household, and both oft hem are meant to—help us keep our households in order, both in society and in nature. As it's currently configured by commercial capitalism, however, the so-called free market is an environmental catastrophe, making money by unmaking the world. It promotes infinite consumption on a finite planet and material production that produces, among other things, pollution, biodiversity loss, and a radically changed climate system that undermines the stability that's essential for good business (among other things). But, with full-cost accounting, the market *can* be configured to operate on the precautionary principle, and to offer people incentives to find their satisfactions in goods and services that are good for the biosphere that nourishes the economy. As an institution that stands partially outside the profit economy, higher education provides both the intellectual capital and economies of scale to change the nature of markets in areas like food and energy.

7. In a culture of remote control, it's time to take control of the supply chains that produce goods and services for us all over the world. Right now, we can take real responsibility for our lives by making sure that the corporations that make things for us do it in socially and environmentally responsible and regenerative ways. In the long run, in a solar-powered world, global food chains and supply chains will disappear except for essentials. So some student efforts can be directed at reviving local production and local markets, cutting the carbon footprint of our food, clothes, cars, and other belongings.[12]

8. Cultures run on peer pressure, and peer pressure is us. Every spoken word and text message, every compliment and complaint, every activity—even inactivity—shapes college culture. Students who conform to the twentieth-century conventions of college culture encourage others to do the same. When they act to embody the new values of an emerging culture of permanence, they exert peer pressure for a more sustainable future. Peer pressure is the power of the people because each student shows every other one the meanings of "normal" and "cool"—which change over time—and which students can change in a more sustainable direction.[13]

---

**The Joneses**: The mythical other, responsible for motivating real people—us—to consume. The figurative family next door who put pressure on us to keep up with their appearance of wealth by buying things. Synonym: *us*.

---

9. We're *all* in this together. Exemplary influence doesn't only influence people nearby. People all over the world are looking to keep up with the Joneses in America—and, whatever their actual name, college students are the Joneses to global youth culture. As such, they have the opportunity to show people a new American Dream based on creative moderation and radical

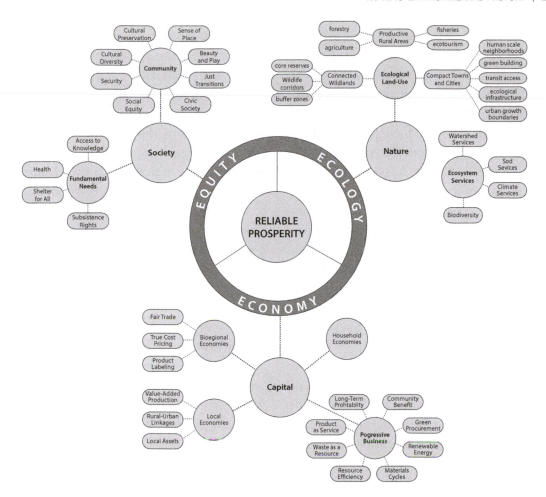

generosity. As citizens, students also have the opportunity to support national policies that show the world that the United States won't continue hogging the world's resources, as well as practical policies that support human development across the globe, so that everybody gets enough of the world's resources. Paraphrasing Martin Luther King, Jr., students can begin to bend the arc of the moral universe so that it tends toward environmental justice.[14]

10. Most importantly, if the pursuit of sustainability isn't also the pursuit of happiness, everyone will end up leading lives of quiet, or even loud, desperation. Environmentalism needs to be fun and fulfilling, honoring the pleasures of the flesh as well as the joys

of conviviality. It needs to remind us of the delights of getting back to nature and of getting back in harmony with nature. It needs to combine the *intensive pleasures* of good food, beautiful clothes, amazing architecture, deep spirituality, and human socializing with the *extensive pleasures* of knowing not only that we've done no harm to the planet, but that we've actually made it better.

This new "commons sense" can sustain an ecological revolution in the twenty-first century—starting here and now on campus. With such mindfulness, college students can transform college culture and the college culture of nature. Through individual actions, peer pressure, and institutional reforms, students can transform environmentalism into everyday life, creating a sort of "*invironmentalism*" as an integral element of who we are and what we do. So far, most college students (and other Americans) have focused mostly on the "mental" part of "environmental." They've thought about their participation in environmental problems, but they have yet to make solutions to those problems part of the pattern of their daily lives. *invironmentalism* needs to be ingrained in language and conversations, work and play, habits, routines, policies, and institutions. *invironmentalism* needs to be "in," and it can't go out of style. Embracing the opportunities posed by our environmental problems, we *can* live happily ever after—which is, after all, the fundamental definition of sustainability.

American college students already make environmental history every day in classrooms and dorm rooms, computer labs and the cafeteria, bathrooms and kitchens. They're making history by their purchases of clothes and computers, iPods and cell phones, TVs and remote controls, cars, food, and fun, as well as with the established practices that govern their use of

such things. Students always make history, but now they have the opportunity to make it by design. And all of us have the chance to shape our lives to shape the future of people and the planet. We have an opportunity to realize our best intentions by devising a culture that uses "commons sense" to solve its environmental problems. This history will also happen in Washington, D.C. and on the world stage, of course. The president and congress might make history by creating policies that make it easier to be good inhabitants of the planet. Other world leaders may do the same. But the ecological revolution of the twenty-first century won't really happen unless we choose to *live* it where we are. It won't be easy, but it could be fun. It won't be smooth, but it will be fulfilling. It will move in fits and starts, and it will be contested fiercely. But that was true of the American Revolution, the abolition of slavery, and the movements for women's rights and civil rights. It's true of most great accomplishments. This challenge is the promise of our lives, of college culture, and—at its best—a new American Dream.

# CREDITS